S0-BAA-251

Dionysos Slain

MARCEL DETIENNE
DIONYSOS SLAIN

Translated by
Mireille Muellner and Leonard Muellner

THE JOHNS HOPKINS UNIVERSITY PRESS
Baltimore and London

This book has been brought to publication with the generous assistance of the Andrew W. Mellon Foundation.

Originally published in 1977 as *Dionysos mis à mort*. © Editions Gallimard 1977. English translation © 1979 by The Johns Hopkins University Press. All rights reserved. No part of this book may be reproduced or transmitted in any form or by any means, electronic or mechanical, including photocopying, recording, xerography, or any information storage and retrieval system, without permission in writing from the publisher.

Manufactured in the United States of America

The Johns Hopkins University Press, Baltimore, Maryland 21218
The Johns Hopkins Press Ltd., London

Library of Congress Catalog Number 78-20518
ISBN 0-8018-2210-6

Library of Congress Cataloging in Publication data will be found on the last printed page of this book.

For Jean-Philippe, Isabelle, and Olivier

Contents

Preface   ix

1   The Greeks Aren't Like the Others   1

2   The Perfumed Panther   20
     The Misfortunes of the Hunt   20
     Le Dit de la Panthère d'Amors   26
     The Wind Rose   40

3   Gnawing His Parents' Heads   53

4   The Orphic Dionysos and Roasted Boiled Meat   68

Notes   95

General Index   119

Index Locorum   127

Preface

Rumor has it that we're not done with the Greeks. Anthropology, if it includes history, is their haven. Why? Because for three centuries they had shared, covertly or not, one kind of knowledge; then, suddenly, up sprang the Greek, Hegelian hero of the phenomenological odyssey, tracing the lofty path from natural to philosophical consciousness. True, this miracle has never been less credible. Yet the subversion of Hellenism is in vain unless it proceeds from within.

In this instance, the Dionysos invoked does not play the role of stranger, still less the role of mute. The byways that link hunting and sexuality, that lead from cannibals near and far to the blood sacrifice cooked by the Titans, traverse a region that is his: one of limits as well as transgressions. We have chosen to interrogate Greek culture at the frontier of its norms and at a distance from the guarantees of humanism that others continue to sign in our name. A system of thought as coherent as the political and religious order of the city is founded on a series of acts of partition whose ambiguity, here as elsewhere, is to open up the terrain of a possible transgression at the very moment when they mark off a limit. To discover the complete horizon of a society's symbolic values, it is also necessary to map out its transgressions, interrogate its deviants, discern phenomena of rejection and refusal, and circumscribe the silent mouths that unlock upon underlying knowledge and the implicit.

Of the two routes to the frontiers—one in the tracks of hunters of ambiguous gender who transgress against prescribed conjugal relations, the other through the labyrinth of dietary customs amid cannibals, vegetarians, and eaters of raw flesh—doubtless the second is today the better marked. It opens upon the reorganized space of mystic belief where the conspiring figures of Pythagoras and Orpheus, ceasing to appear to us as lost outlines and exotic shapes, crystallize into a configuration of the alternative to the city's political and religious system and its worldly order. If the diverse modalities of protest can be apprehended without derision in terms of cuisine, it is simply because the city as a whole identifies itself by the eating of meat—the flesh of a domestic animal cooked on the fire—an act that coincides with the blood sacrifice and founds the dominant values of a world maintained midway between nature and the supernatural. The confines of the sacrifice are a privileged domain in which to follow the paths of Dionysos. First, in the discord he multiplies at will between domestic and wild, men and beasts, gods and mortals, discreet slaughter and violent chase, cooked and raw, and even, according to the Orphic variant, in the culinary process itself, between the spit and the cauldron. The attack on the sacrificial model by Dionysos takes place from without and from within as well. Subsequently, the Dionysiac religion, which even in its mysticism never slips over the brink into absolute renunciation of the world, traces in its travels within the city a route the Cynics took in the fourth century when they set in motion the deconstruction of the dominant anthropological model by aggressive praise of raw eating and familial endocannibalism.

But Dionysos's subversive power is not limited to the frontiers of Greek history. It likewise manifests itself in the heart of the theory of sacrifice that was constituted for us at the end of the nineteenth century amidst the questions produced by the totemic illusion and the reflections of the first sociologists on the interrelation between religion and society. One of the fundamental propositions of *The Elementary Forms of the Religious Life* is that society gains self-consciousness and establishes itself by means of an emblem chosen from among the forms of the animal or vegetal world with which men in a primitive state had the most immediate, even familial rapport. The immolation of the totem animal and the feasting upon it that followed had to be the prototype of all sacrificial practice. A god like Dionysos, who oscillates between beast, plant, and human appearance, straightway found himself at the center of the problems of partition between man and the animal or vegetal world.

Thirsting for the blood of human or animal victims, but in turn himself slaughtered and handed up to be devoured, Dionysos seemed to offer in his ambiguous role of victim and god of mysteries the synthesis of a historical process that began with the savagery of the "peoples of nature" and climaxed in the spiritual maturity of the Christian religion, whose god is a personal one immolated because he sacrifices himself. Such was the strange illusion of a theory that, wanting to distance the menace of a confusion between animal, man, and divinity, beguiled itself into seeking in Dionysos the disquieting precursor of a religious doctrine founded on the concept of sacrifice as limitation of material desire and renunciation willingly accepted by the individual I.

Returning to the Dionysos of the Orphics, those marginal creatures who placed the prestige of the slaughtered god in the service of a radical critique of the city's dietary and sacrificial model, we discover at once how Christianity, through its successive apologias, progressively imposed upon a doctrine that considered itself secular and sociological the gist of its own sacrifice problem. The fascination exercised by Dionysos on the ideologies of sacrifice has no other secret than the ancient connivance of this god with the images of man's delimitation from alternate worlds, gods on one side, brutes on the other.

The route of the hunter, the other pathway to cultural limits, is a detour no less familiar to the liminal god of this book. One of the symptomatic gestures of Dionysos's madness, the *oreibasía*, is to roust married women out of the house on a chase across forests and mountains where every living thing, whether animal or human, is hunted and captured by the savage horde. Likewise, in the order of myth, one of the pertinent traits of Dionysiac transgression consists in substituting the brutal pursuit of animal or human game, which is then torn apart bare-handed, for the nonviolent slaying of a domestic animal, which is eaten cooked at the conclusion of the sacrifice. The hunts preferred by Dionysos derange marital space as well as the orderly sequence of the sacrificial ritual.

In this instance, the panther featured does not at first know Dionysos as master. It belongs to a bestiary in which hunting, seduction, and marriage play shadowlike across the intersecting myths of Atalanta and Adonis. Both are devotees of the hunt. Atalanta flees marriage, refuses the gifts of Aphrodite, and seeks, in a fundamentally masculine and warlike activity, the refuge that will keep her at the greatest distance from amorous desire and the conjugal state. A woman-in-arms, swift of foot and limb, her hatred of marriage is particularly plain in the trial of speed she imposes on her obstinate wooers. Instead of males rivaling one another in ardor to reach at the finish the woman they desire, as is the custom in certain marriage contests, here instead men are forced to flee, naked and defenseless, from the threatening woman they lust after. She chases them before her like timid hares and frightened fawns—precisely the same quarry the hunter Adonis chooses to pursue. In abandoning himself in the woodlands to the excessive passion that binds him to his mistress, Aphrodite's lover excludes himself from the virile world of face-to-face encounters with ferocious beasts. The hunting behavior he chooses, which is the inverse of Atalanta's, leads him to confound the art of tracking game with the art of pleasing and seduction. Adonis's hunt is the continuation of his seduction, using the same means and with the same weapons; his complicity with the panther so testifies, since it is the sole animal endowed with a pleasing natural odor that permits it at one and the same time to captivate and capture its victims.

In this territory, which is forbidden to women and which young men may only pass through as initiates, the huntress who flees marriage meets the hunter who is an effeminate seducer. Their meeting becomes an encounter in which the transformation of Atalanta into a savage beast detested by Aphrodite seems matched by the slaying of Adonis, whose sad end is sadder still for entailing the failure of his proper metamorphosis into a spice plant. At the conclusion of her misadventures and despite the ruses of Aphrodite, Atalanta finds herself confirmed in her vocation for the hunt and her refusal of marriage. She is in effect integrated into the world of ferocious beasts, a world where the invincible enemies are recruited who come to interrupt the career of a young man too much given to confounding hunting with seduction. Nevertheless, Atalanta's transformation into a frigid lioness denounces at the very instant of her apparent triumph the shortcoming with which she is afflicted. It is as though, through these mythical tales and several others related to them, a certain kind of hunting behavior came to mark off a liminal place that was open to transgressions against dominant sexual behavior but could only emphasize the impotence of an efficacious subversion of sexual relations as society prescribed them.

This double domain, at the edges of the sacrifice and at the intersection of hunting and sexuality, is the one I have chosen, primarily because of its richness in forms of transgression but also for two complementary reasons. First, it reveals the effectiveness of both the sacrifice and the hunt as mythic operators, which will uncover, in an elaboration of the preceding analyses, the pertinence of relations and systems of relations that lie beneath the surface of other sequences or episodes of the same mythology. Thus, by way of Atalanta's and Ovid's narrative, we can confirm through symbolic homology the affinities elsewhere demonstrated between the anemone, a flower without fruit or perfume into which Adonis the hunter is transformed, and a plant like lettuce. This reflects on the validity of an interpretation in which only the data, the traditions, and the beliefs of one society can serve as evidence. Such an interpretation should not only be coherent and economical; it ought also to have heuristic value, in such a way as to make apparent relations between heretofore alien elements or to resume information attested explicitly but at some remove in the same system of thought and within the same culture.

We would thereby undertake to respond to objections that have come from the historians of clear thinking, those who willingly reduce history to that which really happened, who oppose an analysis operating by models and proceeding by systematic deduction to the intuitions of common sense, who are scandalized by the intrusion of logic into a form of thought to which it is naturally alien, the fable. A whole system of knowledge that is widely accepted here bases itself on a few postulates: history is an analysis of real change; that which is semantically true for us must be so for another; the only structures of a text are those that can be read at the surface and are formulated in explicit terms. But the true problems for myth analysis are not tied to the illusions of reality propped up by the practice of traditional history; they arise in connection with the reading of a text, with the relations between what a mythical narrative formulates explicitly and the different degrees of implicit knowledge the analyst can and should summon up, depending on whether he chooses to lead his interpretation to the sequestering of a privileged version or, on the contrary, to its opening on the more or less extended horizon of the ensemble of myths a culture has at its disposal.

1

The Greeks
Aren't Like the Others

> If an ancient people
> attained the outer limits
> of civilization, it was
> certainly the Greek
> people.
>
> *A. Lang & Co.*

The relations between Hellenists and those who practice the analysis of myths are not necessarily cordial. It is not, as some might think, only because the former, men of experience who are sobered by the Greco-Roman heritage, are rightly wary of novelty in method. The unease is more serious. Various misunderstandings and several misinterpretations have fed it. "Structuralism" is no doubt responsible for some, while the Hellenists are to blame for the rest. In the intellectual scheme of Claude Lévi-Strauss, Greek mythology finds a natural place beside African or Polynesian myths, since the fundamental goal is to show that mythical thought, wherever it appears, makes reference to what Lévi-Strauss calls a system of axioms and postulates. From this point of view, establishing that Greek myths are amenable to structural analysis means proving that myths uttered two millennia before American myths generate in similar fashion an image of the world long since inscribed in "the architecture of the spirit." But no interpretation of Greece is innocent, least of all when it is motivated by a decision to take account of the way the human mind functions. Was it not Greek society that produced, in the words of Lévi-Strauss, the decisive "bouleversement [overthrow]"[1] that was to permit the emergence of scientific thinking, beginning with the invention of philosophy and the establishment of rational thought?

And so the first misunderstanding breaks out. For Lévi-Strauss, Greece is no more the occupant of a privileged position. It is only the place where mythology gave way in favor of philosophy. The sole merit one can ascribe to it is that it offers an example, doubtless the most complete, of mythological thinking surpassing itself, of that step toward abstraction, which is clearly visible in certain operations undertaken by vast ensembles of myths throughout the world but which knew, in Greece, an apparently happier result. There, by accident, thought in its savage state was transformed step-by-step into a system whose essential principles we still adhere to. Worse still, faced with a theory that expanded the problem zone to the dimensions of the human mind, Hellenists had the sensation of being dispossessed of a brief. To be sure, they tended to discuss the situation in terms of the only split traced by nineteenth-century scholarship, that between *mûthos* and *lógos*; but the issue, they believed, could only be resolved by an exhaustive preliminary analysis of the myths produced by Greek thought or a prolonged confrontation between the forms of thought at work in philosophy, law, or politics and those that organize the whole of mythology throughout the centuries and across the continents.

This first misunderstanding came to be aggravated by another for which structural analysis seems obliged to bear responsibility. In 1955, after having posited that myth is a use of language in the second degree, Lévi-Strauss undertook to define the entity that, faithful to his linguistic model, he then called a mytheme, or constitutive unit of myth. To illustrate his technique of carving up the myth into phrase-relations regrouped according to their thematic affinities, he chose a Greek myth. It happened to be the story of Oedipus.[2] Doubtless the reason advanced was that no one is supposed to be ignorant of this myth—which leads one to ask if its choice was not favored by the cultural status of the Oedipus story in our own society. No other example could more securely offer a magisterial methodology the immediate satisfaction of access to universality. Lévi-Strauss said it himself: Oedipus is the choice of a huckster, "an example treated in an arbitrary way."[3] The principal benefit of the demonstration was to show that the myth should derive its meaning from the relations between its various mythemes. As for the rest, the result was less positive. The totality of versions, which this type of analysis boasts it can take into account to define each myth, was here replaced by a story stripped of the traits pertinent to a structural analysis, since the sequences of the Oedipus myth distributed in his matrix were in fact borrowed from the sociological interpretation that Marie Delcourt had made ten years before without any structuralist perspective. Besides, how could a structuralist analysis work without direct and detailed knowledge of the ethnographic context, whose capital importance Lévi-Strauss was to discover and explicate in his decipherment, during the same year that *Anthropologie structurale* was published, of the "Geste d'Asdiwal"?[4] No Hellenist could side with a method whose only verifiable application clearly revealed that it proceeds with as much arbitrariness as other, more proven, methods, however plainly bedeviled they might be by the discovery of an a priori meaning. The denial that greeted Lévi-Strauss's analysis of the Oedipus myth was even more justifiable, since it could advert to the true principles of structuralist interpretation laid out for the first time in the study of the myth of Asdiwal and published in 1958.

From this reading of the Oedipus myth was born one of the most accepted and widespread of all the misinterpretations that "structuralism" has provoked. Taking as his pretext certain formulations of Lévi-Strauss that seemed to define myth as an effort to find mediation between contradictory terms, the English anthropologist Edmund Leach came to the extravagant conclusion that the mediating aspect of myth was its only function.[5] This functionalist misunderstanding, which made myth a logical tool designed to assure mediation between two contrary terms or situations, led the same anthropologist and several others in his train to propagate a certain number of analyses of Greek or biblical myths whose least deniable originality is to prove that one can call oneself a structuralist while continuing to ignore the procedures and the means elaborated by structural analysis for a decade or more.

These few misunderstandings would not have seemed so profound had they not been exacerbated by the nasty quarrel structuralism started with history and the newest practices of contemporary historians. Convinced that they were forced to choose between "hot" society and "cold" society, between cumulative history and stationary history, Hellenists and several others had no trouble persuading themselves that Greece, which had interiorized its history early on by forging its historical thought in parallel to its political practice, naturally belonged with the "hot" societies that structural anthropology did not seem to appropriate with much self-assurance. The brutal distinction between "hot" and "cold" societies could only accelerate the process of withdrawal already initiated by a series of often unspoken disagreements.

Since the nineteenth century, the great works of classical mythology had taken two postulates as their point of departure: first, that myths are dependent on history, that history is responsible for their chronology and their localizations, and that this history's constant objective is to establish dates, point out inheritances, follow progressive changes, and localize tales by pinpointing the geographical contexts in which they arose and developed. The second postulate, a corollary of the first, is that mythical narratives are composed of a plurality of images and themes whose only relationships are parental and hereditary. For the philologists doing research in this field, the essential problem is, first of all, to identify the original version, which is the only one to preserve the rare perfume of authenticity, then to isolate each of its components in such a way as to determine their respective meanings and, conversely, to show how these different elements have combined and reacted against one another. Yet for more than half a century these two postulates of historical knowledge have not ceased to be reconsidered at the level of new practices brought to bear in a whole series of sectors of historical research, from demographic history to the history of technology by way of the history of mentalities and social relations. Two traits dominate this new historical practice. On the one hand, by working minutely on series of data, historians now take into consideration totalities whose apparently dispersed data depend upon rule-bound transformation systems that are subject to laws. On the other hand, in attacking phenomena that largely elude conscious data, this type of history discovers the importance of centuries-long changes, the long-term ebb and flow, as well as the weight of constants in a domain where movement and change seemed to govern exclusively. It is now almost forty years since Georges Dumézil asked the historians of ancient societies to enthrone structure beside chronology, to establish the analysis of "primary complexes" beside that of "secondary complexes," the distinction being between the social and religious dimensions of thought on the one hand and that which is explained by the successive contributions of history on the other.[6] For all these forty years Dumézil has not ceased inviting historians of religion to recognize the existence of complex states, where, as he puts it, numerous elements are not only juxtaposed but also articulated. And it has taken all this time for people here and there to understand, for instance, that in polytheistic religions the pantheons are not vague, more or less vast collections of gods, but that the analysis of divine powers necessarily coincides with the definition of their differentiated relations within a "structured" ensemble.[7]

At the very end of the nineteenth century, Victor Bérard and a few adventurous comrades were hoping to reintegrate three-quarters of Greek mythology into history, thus saving it from the nonsense that threatened to engulf it. Today historians of the Greco-Roman World seem less optimistic. Some are beginning to intuit that a myth is not necessarily the creation of the historical, geographic, and social milieu in which it appears spontaneously to be situated. Others refuse to question a form of discourse that seems so lacking in coherence, declaring that the only unifying feature of mythology is the provisional and artificial form the authors of mythographic anthologies and manuals gave it in the Hellenistic era. If historians of the Greek world are disappointed, philologists are, for their part, tired of searching for the authentic version that is always simply the one form of a myth that cannot be found.

Forsaken by one group and neglected by the other, mythology can accordingly be identified as tradition: an autonomous class of significations that fundamentally govern the production of each text.[8] It's a sort of postulate of structural analysis; the mythology of a society consists of an ensemble of tales that have more affinities with each other than with any other discourse or form of thought to which the wiles of chronology or the caprices of documentation have associated it. The question is whether the different versions of one and the same myth ought not to be confronted first with one another instead of either being neglected to the benefit of one version or related to data of another class, be it historical, social, or economic. To claim autonomy for the mythological tradition is also to break with the philological habit of parallelisms, of regrouping resemblances and effacing differences by reducing them to fictitious idiosyncracies of the individual imagination. To exploit the peculiar properties of mythology and include the whole ensemble of transmitted or recited myths, it is necessary instead to bring together the different versions by virtue of their differences and to try to see if they cannot then order themselves in the space the mythological tradition opens to them. This is where a structural approach to myth begins.

But before examining the principles of its application, it is necessary to raise the prejudicial objection that this type of analysis can never be applied to the mythology of the Greeks. In an article published in 1972 in the *Rivista storica italiana*,[9] G. S. Kirk, who on occasion does not refuse to treat certain tales in structuralist fashion, insists on what appears to him to be an essential trait of the myths of Greece: they are always defaced by a kind of rereading based on successive interpretations, the most learned of which, by Hellanikos and Pherekydes in the fifth century B.C., are content to reproduce more ancient ones dating from Homeric times. According to Kirk, the Greeks unceasingly manipulated their myths to render them more "credible," either by adding new sequences or by inflecting them as a function of constantly novel local traditions. The result would be that, in contrast to myths transmitted by oral-traditional societies, to which Kirk concedes the privilege of faithfully reflecting social structures, Greek mythology is condemned to appear to us only and ever in a state of incessant agitation and deceptive, secondary elaboration. One may ask if this objection arises from an illusion concerning both the practice of structural analysis and the nature of myths produced by oral-traditional societies as well. In fact, the predominance of the spoken word in a society does not imply its innocence of change. If caught in the trap of a distinction between "cold" societies and "hot" societies, one misunderstands the importance of the continuous warping, reworking, and reinterpreting evidenced by the different versions of a single myth depending on whether one version is told fifty miles away or uttered fifty years sooner than another. Mythology would not be itself without this perpetual reworking of one version by another. For structural analysis, that is precisely the principal justification for its practice: because it has defined myth by recurrence and repetition in variation, it has given itself the task of deciphering simultaneously the different versions of a single myth in such a way as to recognize their hidden system. Consequently, the more numerous the variants, the better structural analysis works. In reality, it would be more pertinent to object that Greek mythology offers more than once the documental aporia of the isolated myth known in a single version and thus despoiled of the essential context a variant could secure for it.

Once the autonomy of mythological tradition has been recognized, the first task of structural analysis is to construct its object. For in principle it does not apply to an isolated myth but to a group of narratives, either the group formed by a myth and its different versions or that comprising two or more different myths; naturally, the two sorts of groups are not mutually exclusive. Very roughly, we can distinguish three kinds of mythological material in Greece. The first—and it is the most important in terms of size—consists of mythographic collections compiled during and after the Hellenistic era. The best known are the *Library* of Pseudo-Apollodorus, the *Fabulae* and *Astronomica* of Hyginus, Book IV of Diodorus's *Historiae*, the collection of *Metamorphoses* by Antoninus Liberalis, and the compilation called *Mythographi Vaticani*. Beyond this technical literature, there are two kinds of isolated mythical narrative. The first kind are those transmitted as fragmentary information by the interstitial literature of the learned, from scholiasts to lexicographers. The others come down to us in a specifically literary (in the Greek sense) work, be it epic, tragedy, or lyric poetry such as the pean, the *thrênos* (lament), or the hymn. Structural analysis claims the right to combine these different mythical narratives and dispose them in the ensembles and groups it is determined to constitute without constraints of space or of time. That is why the first procedure, at least in the case of Greece, is to deconstruct, to shatter the traditional representations borrowed from the apparent sense of the myths, and to hurl down the barriers raised up by ancient mythographers and reinforced by modern mythologists who have faithfully respected their wishes. In principle, each myth refers to an ensemble of other myths. In fact and in practice, the first outline of a grouping can benefit from affinities between personages, actions, or dominant themes. Thus, in a first phase, it is doubtless not useless to confront among themselves the different myths that recount the invention of agricultural life or even to bring together the story of the Danaids and that of the Lemnian women. The danger of such reconnoitering is to confuse structural analysis with the search for a single structure repeating itself in different contexts, such as the motif of the murderous wife or the theme of the simple opposition between savagery and civilized life. The true contents of a myth and thereby its place in a group of narratives only take shape through the decipherment of the different planes of signification without which there is no structural analysis. In fact, structural interpretation is committed to determine the constitutive elements of a system without ever prejudging their signification, which in any case is never contained in an isolated element. Just as a myth can only be defined by the ensemble of its variants disposed in a series to form a group of "permutations," so the same myth can only be demarcated by the different planes of signification that inform it and constitute it as an object of this kind of analysis.

To a certain extent, the Lévi-Straussian analysis of myths develops in the same conditions as the philological and comparative analysis of the nineteenth century. Now and then the point of departure is the same: the gratuitous and nonsensical character of mythical discourse. For Max Müller, the senselessness of myth was scandalous. For Lévi-Strauss, it is a challenge. A structuralist reading begins by positing that the myth is not a succession of words, a story endowed with an ordinary linguistic signification, but a chain of relations, a succession of concepts, a system of signifying oppositions distributed on different planes, at various semantic levels. These Lévi-Strauss calls "codes," because their constitutive units seem to follow the same laws of economy and redundance as true codes. Beyond the sequences that form the apparent content, the analysis therefore discovers, distinguishes, and marks off the different planes of signification that form the architecture of the myth. The weakness of the first structural analysis, the one Lévi-Strauss applied to poor Oedipus, consisted in proceeding to carve up the myth and reorganize it as though it were its own relational context. Since it was defined solely by its own conceptual system, the myth could only be validated by its internal coherence. Thus it became subject to the ingenuity and caprice of the model's constructor. The essential contribution of the "Geste d'Asdiwal" has been to subject formal analysis to the indispensable referent called "ethnographic context," i.e., the totality of information that, for a given society, constitutes the semantic horizons of its mythology, from technological and economic data to beliefs and religious representations, including geographical realities, social structure, and the entire network of institutional practices. The list of the myth's pertinent oppositions at various levels thus attains the indispensable support that only the profound understanding of an organized semantic setting can confer. Structural analysis is not the talky formalism that some accuse of schematism and others of useless complexity.

In order to define the different planes of signification, the analysis should begin by enlarging the field of mythology to include the totality of information about all facets of the social, spiritual, and material life of the human group under consideration. For example, the appearance of the myrrh tree in the Greek myth of Adonis presupposes a scrupulous inventory of all the evidence revealing to us the way the Greeks represented myrrh and spices in their relationship to other kinds of plants. This inventory not only leads to a definition of the use of myrrh in sacrificial practice or the function of perfumes in sexual life but also demands that we pose questions of the ensemble of botanical, medical, ritual, and zoological knowledge through which the Greeks transmit to us their classification systems and whole sections of their symbolic system as well. Lévi-Strauss is the one who has taught mythologists that in order to understand the signification of a plant or an animal, it is necessary each time to determine precisely which role each culture attributes to that plant or animal within a classification system. Nor should one forget that of all the details about such a plant or animal, a given society retains only certain ones in order to assign them a signifying function, and, moreover, that each of these details is able to receive different significations.

Only at the conclusion of this decipherment of the ethnographic context can the analysis determine the conceptual relations whose pertinence has by then been validated by the recurrence of the same semantic values from one end to the other of the domain ruled by the symbolic thought process. To repeat, the group to which a certain myth or narrative belongs becomes apparent only through the planes of signification discovered by this analysis. An example chosen from the Greek corpus permits us to illustrate the interdependence of the operations involved in this decipherment: the Orphic myth of Dionysos slain by the Titans.[10] At once, the story presents a two-fold enigma: on the one hand, it tells of a monstrous meal concocted by

cannibals—the child Dionysos eaten by his enemies—while at the
same time it occurs in the center of the anthropogonic myth of the
disciples of Orpheus, whose thought is utterly dominated by the
refusal to shed blood. Secondly, the cooking technique in which the
Titans indulge is strangely whimsical. They roast their victim's flesh
after first boiling it. Dionysos is transformed into a boiled roast.
Taken by itself, the Orphic myth is a paradoxical discourse that
people have always tried to explain either by trying to see it as a
reflection of a Dionysiac ritual (the *diasparagmós*) or by reducing
it through comparisons to the almost natural representation of a god
who dies and is reborn. Yet a structural analysis can show that the
strangeness of the Orphic tale disappears when it is confronted with
sacrificial procedures, with the fundamental relationship between spit
and cauldron, and beyond that with the ensemble of significations
the Greeks ascribed to roasting as against boiling food. Likewise, by
the specific detail of the gypsum with which the Titans are covered
at the moment they work their violence on Dionysos, the actors of
the myth unmask themselves as primordial men, sprung from gray-
white earth and associated with quicklime. Thus the narrative refers
by its most pertinent traits to an ensemble of mostly mythical representations related to dietary customs, culinary procedures, the blood
sacrifice, and thereby to the human condition as defined in relation
to gods and in relation to beasts. The myth of the murder of Dionysos
turns out to have its place in a series comprising the myths of
Prometheus, the representations of Dionysiac *ōmophagía*, the Pythagorean speculations on the death of the plow ox, but also including
perforce the various tales that the city elaborated in the framework
of the Bouphonia ritual about the slaying of the ox, friend to man.
These tales are linked to another series of myths, for instance the
one centered on the story of the slaying of the first sacrificial animal.
Only in relation to this ensemble, which is recognizable within the
myth of Dionysos, does the Orphic narrative not only take on meaning in each of its peculiar details but also become an element in a
wider system centered on the blood sacrifice.

As for the meaning of the myth, structural analysis no longer seeks
it at the level of plot alone or in the motive forces of the unfolding
story. It finds it at the level of the system formed by a group of
narratives. But even so it does not make, as some have recently
accused, a choice for syntax against semantics. It simply posits that
one arrives at the meaning of myths by multiplying the formal
analyses that permit one to discern the logical armature of several
narratives. The semantics of myths is richer for being discovered
through syntax. And if the method deployed by Lévi-Strauss has

often obscured the semantic values of a group of myths to better clarify their logical preoccupations, one should not hasten to conclude that this type of interpretation is exclusively devoted to the combinatory analysis of X groups of myths. The same practice can lead to the decipherment of a limited group centered on a preferred theme. Pierre Smith and Dan Sperber have shown that, from the viewpoint of structural analysis, myths are not just classifications, a discourse on the logic of propositions whose means and instruments are fire, cooking, animals, and plants; myths are at the same time a body of knowledge about categories and the world.[11] The limited experience of deciphering a certain number of myths, some discussing honey and others spices, allows one to see how much all this unknown mythology—unknown because it is scattered in the interstitial literature—refers to the same body of knowledge about the world and thus to the same system of thought. It is as though a whole section of Greek mythology, while discovering the horizons of agricultural life, was only exploring a single, selfsame problem, namely, the human condition with respect to marriage and sacrifice, the two institutions that support a whole section of Greek symbolic thought.[12]

Even the most intractable Hellenists will no doubt concede to structural analysis that it can contribute to the inventory of mythology's riches, but they will insist on making two major objections to its application in the domain of Greek myths. The first is that this kind of analysis cares little about the particular forms a mythical narrative may take, whether it is the skeletal résumé given by a late mythographer or the version deployed in a tragedy of Aeschylus. The second objection is that structural analysis ignores factual data and seems to belittle the importance of social realities. Both difficulties pose essential problems for this kind of analysis in Greek studies.

From the linguistic point of view, Lévi-Straussian analysis proceeds in a slightly paradoxical fashion. On the one hand, it supposes that a myth does not coincide with language data, with the phrases of a narrative, and on the other, it makes a series of loans from structural linguistics in order to apply its conceptual apparatus to the metalanguage of myth. In fact—we now see it more clearly—the analysis of myths only owes linguistics some of its motivation, and the original procedures it deploys have no more to do with phonological structuralism than with generative grammar, which the combinatory analysis in Lévi-Strauss's *Mythologiques* sometimes seems to reflect.[13] At the conclusion of the investigation opened by *The Raw and the Cooked*, the project of making a grammar of myths, be it structural or generative, runs up against the impossibility of defining the constitutive units of myth. The mythemes remain undiscoverable.

Even so, for the Hellenist mythologist, confronted by a series of mythical narratives he has long since learned to read as "texts," the problem of linguistic structures and their relationship with the structures of myth remains an essential problem. It is pressing insofar as in this mythologist's domain a great number of variants take the form of texts as coherent and elaborate as the epinician odes of Pindar or the tragedies of Sophocles. Schematically, one can distinguish three situations. In the first, the narration is reduced to a minimum: these are the résumés of myths that are reduced by mythographer after mythographer until they are nothing more than a juxtaposition of impoverished sequences that offer no resistance to structural analysis (which is, moreover, often condemned for contenting itself with the oppositions left by this sclerosis of the narrative). In the second case, the myth becomes a narrative whose logic governs the entire organization of the work. Structural analysis must then be accompanied by a decipherment of the shapes and effects that the play of narrative forms has traced in the very texture of the myth. Principally in question are narratives in verse embedded in such works as the *Homeric Hymns* or Hesiod's *Theogony*. Philologists —for example, Hans Schwabl on the *Theogony*[14]—have already shown that the coherence of these texts is guaranteed by a whole series of syntactic parallelisms. Even if Hellenists have not always perceived that "relations of equivalence and formal parallelisms" impose on the reader or hearer semantic connections between the elements of a narrative and thus create a veritable network of associations, the analysis of the myth can in no way divest itself of them. As for the third situation, it is brought about by such works as choral lyric and tragedy, whose composition obeys certain very strict rules that cannot be confused with those of the mythical narrative. In this context in particular the problem of the *reinterpretation* of myths evoked by Kirk arises. Traditionally, the praise of a victor in the games comprises a mythical narrative. Through it a poet, invested in the case of Pindar with the status of "master of truth," defines the norm of the system of aristocratic values whose indispensable seal he confers by virtue of his praising function. Charged with exemplary value, the myth told by Pindar is thus placed in a new setting: the point where past and present intersect, where tradition and reality combine—though always for the specific ends that the gnomic sentence placed at the poem's conclusion unmasks. In Pindar's case, the reinterpretation of the mythical narrative is sometimes so clear-cut and distinct that it is presented in the form of a correction supplied to the narrative either in an entire sequence or a particular detail. Yet insofar as it is the object of an explicit avowal, the distortion imposed on the myth does not disqualify Pindar's version in relation to others. In a lyric composition such as this, if the myth is semantically renewed, it is only with respect to the exemplary value the epinician ode assigns it.

In reality, tragedy alone threatens myth with a reorganization profound enough to attack, in certain cases, the actual motive forces of myth. The tragic composition is inseparable from mythology. That much is obvious. The myth grants tragedy its essential personages and the great themes of its action. But once taken up and accepted by tragic representation, the mythical story is at the same time—as Jean-Pierre Vernant and others have stressed[15]—distanced. Henceforward the myth is before the eye of politics. The ancient values that mythology brings to bear are confronted with those that the city is busy constructing and whose antagonistic spokesman is embodied in the chorus. Consequently, tragedy utilizes a mythical story through which it puts into question the acts and words of the hero and the actors by constantly passing from the value system of the city to the form of its mythic past. But this critical activity not only touches the succession of actions told by the myth; it also attacks more directly certain essential mechanisms of the myth, particularly when the tragedy undertakes to articulate the shapes of ambiguity and to explore them systematically. Here the mythologist, instead of treating the tragic version of a myth as he would a narrative provided by some poor mythographer, doubtless should precede his analysis of the myth by an analysis of the corresponding tragedy, so as to disclose the modes of reinterpretation and the forms of distortion that are specifically tragic.

The second problem raised by the objections of Hellenists is nothing other than the relationship between myths and, on the one hand, social reality, on the other, factual reality—the clashes, jolts and changes of history. Certain historians of antiquity, such as G. S. Kirk following Victor Bérard, are convinced that it suffices to know the conditions of the first expression of a myth to establish its exact meaning. Once brought back to its origin and restored to its first landscape in history, the mythical narrative ceases to be opaque and sybilline but at once becomes transparent. Such is the method of those who were once called "analysts"; at least, that is what they have retained from the enterprise of Karl-Otfried Müller and his *Prolegomena to a Scientific Understanding of Mythology*, so new in 1825: recover the various circumstances that gave birth to a myth. From this point of view, these "analysts" are inclined to think that Greece is particularly disfavored in comparison to the Americas, and they take pleasure in flattering the good fortune of ethnologists who live amid a population of narratives that speak naively and spontaneously of the society that produced them and whose myths confide without mystery their cares, their problems, and their various anxieties. The poor Hellenists are accordingly forever condemned to know only myths deprived of reference to their original context, myths coming from the depths of the neolithic period or products of the Mycenean world, myths separated forever from that which gives them a meaning, mute myths or ones that speak an abolished tongue that none can dream to decipher. Perhaps we may offer some solace to the historians whose interpreter G. S. Kirk considers himself to be by reminding them that the Americanists are not always in the lee of history and that even in their domain there is a whole series of studies inspired by the Finnish School that struggles courageously to contrive a kind of natural history of the tales and narratives transmitted by oral tradition. Is not their major ambition to show "where they were born, at what period and in what form," in order to classify the variants according to their place and order of appearance?[16]

Let us examine the problem in another way. It sometimes happens, despite the obstacles that some people not unreasonably count upon, that historians succeed in recovering, in the attics of history, the document that seems to record the birth of a myth. Let us take as an example one of those privileged cases that gives the historian of the Greek world the satisfaction of passing in review the different interpretations proposed by several generations of mythologists only to see them swoon before the obvious simplicity of the explanation that History of necessity grants once it consents to address us. In this case the lucky hunter is named Paul Faure, and the myth that benefits from his discovery is the story of the daughters of Danaos, the Danaids who were pursued by the fifty sons of Aegyptos from the banks of the Nile to the springs of Argos.[17] In 1964, in ancient Egyptian Thebes, archaeologists discovered a hieroglyphic inscription engraved for Amenophis III, ca. 1380 B.C. This inscription provides us with a list of names on either side of two prisoners chained to the Pharaoh's foot. One of the prisoners represents the island of Crete, the other the Danaoi, and in the list of towns and districts associated with the Danaoi, the people of Danaos, is to be read the name of Nauplia, the name of a city founded, according to Pausanias, by the Egyptians who came with Danaos (Paus. 4, 35, 2). The royal inscription discovered in 1964 thus provides, according to Paul Faure, the key to the myth of the Danaids: the tale of the adventures of Danaos's daughters and of their Egyptian cousins only relates to us a historical fact whose date is approximately known (ca. 1380 B.C.). We have reached the source of the legend of the Danaids: a triumph of *pegomania.*

Let us admit that the myth of the Danaids was born there on the river's edge where some Egyptians met some Greeks. How does this event explain or account for the essential traits of the myth? Paul Faure does not tell us how far the event goes, nor what relation it maintains with the flight of the Danaids, their refusal to marry their cousins, their race to Argos, where the subterranean waterways dry up as a consequence of a quarrel between Hera and Poseidon. And no one would dare to pretend that a fuller chronicle, with details of marital practices around the year 1380 B.C., would explain to us why the Danaids are at once women who flee marriage with men who are too close to them and wives who are said to have introduced into Greece the great marriage ritual, the festival of the Thesmophoria. To understand this story it is necessary to enter into its unfolding, to follow its sequences, to try to understand the connections between the spring Amymone and the marsh of Lerne, which is also a spring; it is necessary to study the rituals of Argos, which prescribe that the fiancées of a given year draw pure water for the nuptial bath at the spring Amymone; it is also necessary to take on the mysteries of Lerne, the role of Demeter in these mysteries, and all the symbolism of water, water that comes from the subterranean world, that springs from below or disappears in the earth's depths, to establish the relations between disappeared water, sought-after water, water that is carried, whether lustral or defiled, and the ritual recipients of the *loutrophóroi,* the *hydrophóroi,* and the *thesmophóroi.* Such is the ensemble of the ethnographic context of the myth, and it utterly escapes the interpretation that is caught in the trap of the original "historical" reference.

In fact, the event of 1380 can explain nothing. It is a document of political history. Uncontestably, it proves that in the fourteenth century B.C. there were contacts between Egyptians and Greeks in proximity to Argos. But if we had a detailed chronicle of the events that transpired at that time, the myth would be no less opaque. What must be repudiated in the "historicizing" interpretation—which is preferred, consciously or not, by so many Hellenists—is the irrepressible illusion that the discourse of myth must of necessity reflect "reality." The fundamental postulate of the majority of historicist interpretations is the belief that the relation of myths with social organization, with the physical world, with the natural world, and with events is always exclusively of the order of *representation.* Since the "Geste d'Asdiwal," it must be recalled, Lévi-Strauss has shown in a series of decisive examples that mythology could not serve to paint a faithful picture of ethnographic reality and that a myth should no longer be confused with a documentary source whence the ethnologist could fetch what he needed to reconstruct the social organization, the beliefs, and the practices of a society. In principle one can never deduce the real from a mythical narrative. Structural analysis therefore refuses to admit something that not only historians but also sociologists willingly agreed upon in their research, namely, that mythical thought has a *direct* relationship to social data (explicitly so stated by sociologists such as Granet and Gernet) and that certain mythical images can provide the gaze of a trained eye with "prehistoric" behavior and abolished institutions that luminesce only in the light of very ancient ways of thought.

On the contrary, structural analysis has shown more than once, by making reference to the ethnographic context, that the institutions described in myths can be the negation of real institutions or that the animals appearing in mythical narratives behave altogether differently than observation and zoology predict. To be sure, myth entertains a relationship to the environment, to the ecological or social given, to the history of a group; but it is always an indirect, mediated relationship, one that suits an autonomous discourse extracting from reality the elements over which it maintains sovereignty. On one plane of signification a myth can utilize a certain amount of raw data (for example, elements of physical geography); then, on another plane, it can mingle real and imaginary; finally, on a third level, it can reverse, systematically or not, the real data.[18]

History, whether social or narrative history, possesses no privileges for the explanation of myths. It is only one among many sources of information that make up the reality that mythology exploits. It is possible that the event of 1380 had some relation to the myth of the Danaids. Perhaps it was its point of departure. But whether or not it gave the myth its impetus, the event has been devoured by the myth. Mythology has incorporated this fragment of history and rearticulated it in terms of its own proper structures.

We can observe this task of reorganization in certain cases where structural interpretation has been carefully carried out and sent back more than once to the drawing board. The Hesiodic myth of the ages offers a convincing example. Jean-Pierre Vernant's analysis uncovers in the story told by Hesiod not five races in chronological succession following an order of progressive decline, but a three-tiered construction, with each tier divided into two complementary and opposed aspects. This tripartite structure functions on a series of planes that the myth superimposes on the narrative, but it is itself overdetermined by another whose two terms, by virtue of their relationship, give maximal polarity to the construction as a whole. This is the opposition between the two terms *díke* (justice) and *húbris* (excess), whose importance Hesiod himself points to when he draws the lesson for the hearer of the myth that he must heed justice, *díke*, and keep excess, *húbris*, from swelling. But the *Works and Days* does not simply provide us with Hesiod's avowed intentions. It develops before our eyes the socioeconomic context that is inseparable from the myth of the ages.[19] The economic and social situation of Ascra in Boeotia is marked by the subdivision of lands, the dispossession of small landowners, and the progressive indebtedness of peasants. In short, there is an agrarian crisis whose typical aspects—misery, hunger, scarcity of land—Hesiod sets forth but which he interprets with his own catergories, those of a theologian and a moralist. The crisis we call "agrarian" is lived by the poet as an evil, an excess: *húbris* swells and threatens *díke* with extinction. These are the two principles that Hesiod puts at the center of the myth of the ages when he is constrained by certain social and economic circumstances to tell, in his turn, this tale. In so doing he reinterprets it, reorganizes it in his own way, which is the way of his time and history, a harsh, dry, and windswept history like the land of Ascra, "a town accursed, mean in winter, harsh in summer."

In Greece no less than elsewhere, myths are perpetually retouched, rearranged, revised, and corrected. Events, encounters, contacts between groups and societies are all data that constitute the environment as much as the fauna or the climate. Each innovation, each change in the society and its institutions and in social relations can be translated into a reorganization one of whose major functions is to deaden the blows.

Beyond the difficulties discerned by some Hellenists for the proper use of structural analysis in the Greek world, there remain the implicit objections. Immediate evidence for their existence transforms them into presuppositions, or, alternatively, their naiveté today makes them the object of shamed silence. In this land of the unsaid determined resistance is entrenched and rejections or exclusions are legislated unilaterally. One of the most widely shared presuppositions among Hellenists has been hammered into the minds of historians of Greek antiquity: the certitude nourished by the nineteenth century that Greece, more than any other society, is indelibly marked by history, a history that it completely internalized and which thus became an essential feature of the Greek man's experience. No one can deny the importance of Thucydides or of a work in which the recognition of the Greek citizen's historical condition goes hand-in-hand with the fully developed conviction that this kind of man can act in the present and get a grip on the future. This conception of a historical work is central for the meaning of political action in the Greek world. But the *Peloponessian War* has never been a sufficient reason to subordinate all the forms of thought in Greece to the conception Thucydides devised of history, of temporality, and of the action of men. And that certain Hellenists found glory in the exhaustive study of the chronology of the archons is not a better reason for believing, in their wake, that at other levels of thought the activity of the mind is reduced to the same chronological necessity.

At the risk of scandalizing some historians who are jealous of their time-honored privilege, it should be remembered that structuralism is principally concerned with variations and that "change is a particular mode of variation."[20] In this way, structural analysis can, without pretensions to grandeur, make different ensembles appear as variants of one another. In the Greek world, political myths, foundation myths of neighboring or rival cities, should be handled by this method, as should myths about the formation of cults or sanctuaries whose ownership is claimed by several social groups. Most often the claims are made by competing myths, and Greek historians have more than once recognized the affinities and made plain the divergences among them. Furthermore, structural analysis can take responsibility for problems that seemed until now strictly the property of history: for example, the analysis of successive states that are the different forms assumed by a single ensemble. Taking the analyses of Greek mysticism by D. Sabbatucci as a starting point, one can propose an application of this method to organize the system of representations worked out by the Greeks between the sixth and fourth centuries B.C. for the consumption of human flesh and the problem of a meat-eating diet.[21] With respect to the sacrificial and nutritional model dominated by the relations between three terms—gods above, animals below, and men in between—four forms of protest against the City (Pythagoreanism, Orphism, the Dionysiac religion, and Cynicism) dispose themselves by twos in relation to each other, depending on which orientation is chosen. From one perspective, the protest entails transcendence by the upper route (Pythagoreanism and Orphism); from the other, transcendence by the lower (Dionysiac religion, Cynicism). The contrast between the two solutions is epitomized by the representation of extreme cannibalism jointly held by Pythagoreans and Cynics: a child devouring his own father and mother, which is a horrific vision for Pythagoras's disciples of the bestial, carnivorous life lead by Others, but for the Cynics an exemplary image of the radical deconstruction of society that their daily regime aims to realize. Now in this case history teaches us how under certain political, economic, and religious circumstances the metamorphosis of certain Pythagoreans into disciples of Diogenes actually took place. Such a transformation unfolds at a dramatic time, amid failure and violence, but from the viewpoint of the system it translates into the passage from one extreme to the other, a passage that can be designated unemotionally as a positive sign being replaced by a negative one. Thus the system of structural transformations happens to be spread out in time and space, and the structure is simply "the historically real transformation rule."[22] That the example is not isolated does not mean that all change in history is appropriate to this kind of analysis. At least one can allow that in various sectors structural analysis can show how the unfolding of history is at times subject to certain constraints, even if these are in turn dependent on another history that is all-encompassing, void of events, consisting of slow epochs and lengthy traverses.

But the brand of history that Greece bears on its brow would lack the prestige it possesses were it not compelled to confirm the elect character of the Greek people. From Herder and Winckelmann to the nostalgic humanists, the ideology of Greece has never ceased to renew itself in the form of several basic themes each of which pivots on the privileges of the Hellene. Bearer of civilization, discoverer of Reason, the Greek has been singled out from among all peoples to create the Beautiful with the resources of the East and to make a decisive contribution to the progress of human sensibility. From the Jesuits' accounts of the New World holding up a mirror to Plutarch to the declarations of Nietzsche on the naiveté of the Greeks, whose literary works serve us as inaccessible norms and models, the complex connections our societies continue to develop with Greco-Roman civilization urgently demand the *archaeology* of the Western World undertaken by Michel de Certeau and a few others. Today, apart from some decrepit attempts to recall the priority of the Greeks in one domain or the other of contemporary life, the professions of faith in the Greek model are becoming rare and discreet. The candor and simplicity manifested by Brian Vickers are only the more prized as a result. In a new and important work devoted to the problems posed by sociological interpretation of Greek tragedy,[23] the British critic, who brings a vicious suit against his compatriot Kirk for collaboration with the structuralist usurper, proclaims, at the conclusion of a radical critique of every attempt to apply "the structural method" to the myths of Greece, the profound reasons for his position. The position, it turns out, owes the better part of its insolence to the status of "humanist" to which its author pretends, liberated as he is from all the technicalities of this field of study. It matters little that Vickers reproaches Lévi-Strauss for distinguishing in a myth the reality of the concepts from the appearance of the narrative, or even that he charges him with ignorance of Boas's clairvoyant discovery of the fundamental continuity between mythological discourse and social life. What seems more basic is his claim to have direct access to the content of myth: Brian Vickers demands the right to hear the mythological word without being forced to ask its meaning from an analysis that is always worried about what the apparent discourse obstinately refuses to say. Is this the impatience of a reader who must have nothing between himself and the text, or is it the reasoning of a theoretician convinced that if the myths must be decoded, they therefore no longer speak of the society they address? Vickers categorically refuses to hide anything, especially such good, solid reasons. First, the Greek myths are still our own, and since the West has grown up on this mythology, we cannot mistake its meaning; then, in the myths of Greece, apart from a few irrational elements, the subject is man—the mythology of the Hellene is dominated by anthropomorphism. Besides, look at what they invented: literature, art, the sciences, law, politics . . . Nothing to do with the Tsimshian, those salmon-fishermen. No, truly, the Greeks aren't like the others.

2

The Perfumed Panther

De la pantera
Cristo è la fera co lo dolçe odore,
quelle ke corrono l'anime sante,
de le quali per vivo amor se pasce.

Bestiario moralizzato di Gubbio

The Misfortunes of the Hunt

For contemporary historiography, the strangest characteristic of the historian of antiquity is his passion for *autopsy*[1]—not, to be sure, the autopsy of gloved, white-cloaked acolytes whom a police inquest obliges to inspect the corpus delicti, but that of a witness who is all eyes. "If it were possible to be present in person at all the events, that would be by far the best basis for knowledge."[2] For Ephoros of Cumae, writing one of the first universal histories in the fourth century B.C., the unreal modality of words constitutes an admission that exile to the present, to the moment when absence already seems to submerge the greater part of reality, simply inflames the desire for knowledge based on direct, ocular vision of human actions. In a tradition respectful of its Father and convinced that history "has basically remained unchanged" since Herodotus,[3] some contemporaries believe their purview is defined by the testimony of the one who knows for having seen, revealed by the blinding *epoptía* that actual presence at an event confers. Thus for one Hellenist, a scholar of epigraphy, an inscription from Attica to Bactria is a direct witness: "Beneath the sooty water that slightly displaces the finger on an almost defaced piece of marble, a town thought to be 150 kilometers distant appears in an inscription of Ionia."[4] Not only do the stones speak—and their voices do not issue from the farther side of death to which their cold, opaque mask consigns them—they bring into view the spectacle of the past, composing the kaleidoscope[5] by virtue of which the historian/antiquarian finds himself transported to the ancient world: he becomes the contemporary of stonecutter, magistrate, benefactor. . . .[6] History is no longer a matter of aligning or interpreting the fleeting traces of a lost person. The "Fountain of

Youth"[7] gushing up from the epigrapher's stone confers on the historian the privilege of contemplating events directly and with his own eyes. *Epoptía*, moreover, is not solely restricted to the man who deciphers the primary document: "Every reader of the inscription can sense it in turn; once the document is printed on paper it is as direct as when it was inscribed on stone."[8]

In this methodological discourse contributed in 1962 to a collective volume on history and its methods, a certain kind of "realism" reaches its apogee. The narration of past events fades away before the shared vision of a world, thought lost, that rises to the surface to be read in words traced upon marble. It is true that an epigraphic document is not a *monument* among others disposed along the paths that history follows. With its mass and its substance, which impose the real presence of the past, how could it not give a historian the startling impression of being present in person at all the events, thus extricating him from "endless discussions of texts that have been worked over for four centuries"?[9] The only specter that sometimes troubles this solid good fortune is the scrupulously exorcised one of "wandering stones,"[10] rolling across the world, leaping from one place to another. By separating the document/witness from its site and natural geography, it alone can menace the epigrapher's privilege of "making history gush forth."[11]

The illusion of reality does not often have this much force or luster, but in more or less insidious ways the realism of "it actually happened" does not cease to govern the joint understanding of historians who nowadays wish or feel themselves to be "antiquarians." From the viewpoint of visionary "realism," a project like the structural analysis of myths can only arouse suspicion. On this terrain is not the historian deprived of recourse to a continuous chronology, a chronology that fails at organizing myths just as it succeeds in ordering the sequence of archons? Since the heirs of K.-O. Müller discovered that a myth is not necessarily the product of the historical, social, and geographic soil from which it seems to spring spontaneously, mythology, once abandoned, seems to possess only dubious reality. It is neither fish nor fowl, and writing, though severing it from orality, did not thereby purify its tainted relationship to the imagination. To the instinctive mistrust that the historian or prose writer since Hekataios of Miletos or Herodotus developed in the face of myth has been added the positivist anxiety that the "historical" nature of myth sires in societies dominated by writing—its reality in relation to events as well as its relation to a given social, political, or even economic framework. The suspicion that "realist" history develops, consciously or not, with respect to the figure of myth naturally hardens in response to a kind of analysis whose practice leads it to value latent conceptual relations and to reject from the beginning the explicit sense of a tale. Thus it is not surprising that the principal objections addressed to an essay on the mythology of spices in Greece[12] cluster around "historical realities" at times neglected, at times utterly disregarded. In a series of remarks that invite dialogue, Pierre Lévêque regrets that the decipherment of the various planes of signification was not carried out "in a framework more concerned with historical realities."[13] In a long Roman critique whose violence culminates in the hurling of anathema, Giulia Piccaluga, who appeals to an interpretation based on "the historicoreligious method,"[14] denounces the failure of the structuralist enterprise, which systematically falsifies the relations between myth and history. The convergence of these two critiques of differing intent and inspiration indicates the importance of the problem: at stake is nothing other than the nodal relation between Demeter and Adonis at all levels and on the various planes of the mythical ensemble under analysis. For Pierre Lévêque, who is attentive to the historical process of the genesis of cult, the Thesmophoria, a Demeter festival, has every chance of dating from the second millennium, while obviously Adonis the Semite is only a latecomer to the Greek household.[15] The wedge that chronology drives between the authochthonous Greek divinity and the foreign god of the Orient would render it historically impossible that the opposition between Adonis and Demeter designate a fundamental structure in the mythology of the Greeks. Two terms of unequal age could only constitute a poor pairing.

For Giulia Piccaluga, whose system of reference is much grander, the structuralist analysis disregards an essential historical reality: that in the course of millennia, for the most diverse cultures, and in Greece as elsewhere, cereal farming of the Demetrian order can only be established once the civilization of hunter-gatherers has disappeared.[16] In other words, the hunter Adonis belongs to a history rendered intelligible by the emergence of a cereal-based economy. Both critiques choose historical reality as their frame of reference. But, be it chance irony or a ruse of history, Piccaluga's major objection is the inverse of the one formulated by Pierre Lévêque: one reproaches me for failing to realize that Demeter came before Adonis, while the other blames me for not knowing that the goddess of grain came after Adonis. The contradiction between these two historians does not discredit the acknowledged value of chronology. At most it is an invitation to discuss the respective presuppositions behind their recourse to diachrony and, as well, the appeal, which is in both cases implicit, to the reality of history.

In the field of interpretation, one of the sacred cows of historical thought—for more than a century a divided branch of knowledge—considered it plain that mythology is to be explained solely in terms of its past. Behind each mythical shape there lurked a "historical" event that awaited in silence its summons to grant irrefutable meaning to the meaningless babble of the fabulous tale. Depending on whether the founding event belonged to a more or less distant past, the interpreter/historian could gauge the manifestations of reality that guaranteed his tale's credibility. Consequently, long before reaching the borders of the Paleolithic Age, motivations freed from the constraints of erudition invade the interpretive model and abandon it utterly to an exegesis that is internal to the very symbolic system from which the interpretation wished to distance itself. As everyone knows, insofar as archaeology can discover, the first Greek settlements had their material basis in the culture of cereals and domestic animals. Hunter-gatherers are confined to the Paleolithic haze. They are not the less haunting in the interpretations of some moderns whose way of writing history drifts imperceptibly into fabulous tale under the influence of a desire to exhibit origins without even renouncing the demands of realism that constitute the sole horizon of probability.[17] At precisely this point, two theories connive to intersect: the first legislates on the obligatory retreat of the hunter-gatherers before the new wave of grain eaters; the second is convinced that the Paleolithic Age is the only one long enough to account for the genetic evolution of humanity and that hunters during its several millennia were responsible for inventing the gestures and behavior that so marked mens' memories and that mythical narratives, replacing long-since enfeebled rituals, sought to resuscitate.[18] Without a psychic archetype or some antievolutionary structure, it is impossible to conceptualize the workings of this Paleolithic heritage, a heritage for which the Greeks themselves are invoked as more or less conscious witnesses by virtue of an isolated ritual or some unique tradition in their history. In the contemporary historian developing such ideas there are doubtless motives that could shed much light on our own "mythology" about hunting and on the place our ideology accords to this activity. Unfortunately, one must invoke the sluggish, crooked progress of history, which some analysts of ancient myths deplore in structural interpretation, to account for ignorance of the historical reality that the same scholars are striving to reconstruct in the conviction that only the past can serve as a guide to mythological tales. But, clearly, is not one of the first problems with the process that these phantom hunters from the Paleolithic Age render vain and futile all our information from the first millennium B.C. on the socio-

economic status of hunting? For hunting, in contrast to tribal initiation, is neither a residual element nor a moribund institution, from
the time of the collective expeditions of youths in epic ideology to
the tracking of hares by stout squires in the age of Xenophon, an
author whose *Art of the Hunt* seeks to remind his contemporaries of
proper etiquette in this domain. Between the fifth and fourth centuries
B.C., during the period contemporary to the Athenian tradition of the
story of Adonis, Greece keeps in active service, side by side, the
citizen-hunter and the Spartan *kruptós*: the *petit bourgeois* who leaves
behind the turmoil of the assembly for a day in the country, and the
frightening nocturnal man, as dangerous as a wild animal, for whom
hunting is part of the hazing process, part of a vast initiatory complex with certain formal correspondences to initiation practices in
"primitive" societies, as A. Brelich has recently pointed out.[19]

To discount contemporary history in favor of a distant, almost inaccessible past, it is still necessary to be convinced that one should
and can reconstruct through representations and interwoven ideologies a kind of socioeconomic "warp" of hunting that will make plain,
by way of contrast, the design of the mythical chain. No serious
argument sanctions the belief that mythology is a waste product or
by-product of history. To the contrary, certain analyses and theoret-
ical reflections concerning myth suggest that the different planes of
signification as they traverse the whole of mythology have a broad
autonomy at their disposal. If, for example, hunting orients a series
of myths in a society as fundamentally agricultural as Greece in the
first millennium B.C., it is not a distant but a faithful echo of the
social relations of production for a horde of hunters who crossed the
clearings of history several millennia before. Or, more concisely, it
is because, in the order of myth, the activity of hunting constitutes
an excellent operator. And this for a series of reasons: as a fundamentally masculine activity in which confrontation with wild animals
leads to the spilling of blood at the same time as it provides meat as
food, hunting contrasts with farming but is closely linked to war.
And if war, as a mortal occupation, is exclusively the male's preroga-
tive, the production of food through farming, on the contrary, works
on the model of gestation and reproduction, even if in Greece men
are the ones who till the earth. Situated at the intersection of the
powers of life and the forces of death, the hunter's space constitutes
at once that which is beyond the farmer's fields and their negation.
Choice haunt of the powers of savagery, the domain open to the
hunter belongs exclusively to the male sex. For the young boys who
venture within it, alone or with their age-mates, a hunting exploit
assures their integration into the political class of adults. In traversing

these wild places the male child, snatched from the shade and warmth of female bodies, is introduced to the realm of manhood. By confronting wild animals, he prepares himself more or less directly to become a warrior initiated into the unshared privilege of men: violent bloodshed. Artemis, virgin mistress of the hunt, only half-opens her domain of woodlands and mountains to the young girls consigned to marriage. As little "bears," they cannot even leave the enclosure of her sanctuary. There, in the very spot where a man murdered a household she-bear of Artemis, the little girls of Athens are obliged to expiate his crime by entering the service of the goddess and "doing the bear" in saffron robes, obtaining, at the conclusion of their novitiate, the right to abandon their virginity, reenter the city, and become wives and mothers.[20]

Forbidden to girls and traversed by boys before they accede to the status of warriors and adults, the hunter's terrain is not simply the negation of the farmer's fields and of the enclosed space of the home. It also constitutes a space outside of marriage that welcomes deviant forms of sexuality or those that are simply considered strange by the city-state. Thus a system of relations seems to form between hunting and sex. Out of hatred for women, a young man goes off to track hares in the mountains and never returns. To flee the marriage that stalks her, a young girl decides to run away and make war on wild creatures at the mountains' lofty summits. But in masculine love affairs, the gifts given by *erastés* (lover) to *erómenos* (beloved) are often the fruits of the hunt (hares, deer, foxes).[21] Similarly, in the Cretan courtly tradition, it is during the two-month hunting season that the adolescent boy is customarily stolen away to have intimate relations with his male lover before being, in this case, integrated into the brotherhood of warriors. Access is granted by the ravisher himself, when, at the conclusion of the initiation, the beloved obtains his war dress from the very hands of his lover.[22]

Forests and mountains compose a masculine landscape from which the woman/wife is radically absent; so, too, are excluded the sociopolitical values that define the proper use of the female body. In the space where social rules are silent, deviance is articulate, and transgressions come to pass. There Hippolytos, madly in love with Artemis, the virgin's inseparable companion,[23] and a youth consumed by the fire of continence, has the daughters of King Proitos as his neighbors, women crazed with desire and utterly nude since the day they offended Hera; Dionysos, savage hunter, leads in his train the pack of wild women who have left behind the loom, abandoned their husbands and hearths;[24] Atalanta, driven by the odium of Aphrodite, pursues with javelin in hand the dream of immaculate virginity, while the "multiple-murderess," *Poluphontê*, having left home to "do the baby she-bear" and live in sole company with Artemis, falls madly in love with a real bear with which she makes love beneath the horrified gaze of the virgin huntress herself.[25]

As a result of its position between war and marriage, the hunter's terrain gains its capacity for becoming the privileged place in myth for marginal sexual behavior, whether it be masculine or feminine denial of marriage or, inversely, experimentation with censured sexual behavior. As a liminal place where socially dominant sexual relations are as if suspended, the land of the hunt is open to the subversion of amorous pursuits, whatever their process or modality.

Le Dit de la Panthère d'Amors

In the perspective opened at the intersection of hunting and sex, the example of Adonis should serve as a touchstone. At the very least it leads one to reexamine a hunting act whose deadly consequence polarizes attention and tends to orient interpretation in one direction only. In a previous analysis, I insisted on the negative aspect of the confrontation with the boar, showing how the tradition of the fourth through third century B.C., from the comic poet Euboulos to Nicander of Colophon, stressed the flight before the monster's charge and the folly of a hunter whom fear forces to take refuge in a bed of lettuce, unless it was his mistress who was so stupid as to offer him the protection of a plant whose deadly virtues she was destined to learn at her own cost.[26] The impotence that results impugns, on the erotic plane, the virility of a hunter who has neither the courage to dispatch a boar with the lance nor to await steadfastly his onslaught, thus proving he possesses neither the valor nor the warrior's fury of his adversary. But Aristarkhos is not wrong: it would be as ridiculous to confront the adventure of Adonis with the exploits of Herakles, the killer of monsters and destroyer of wild animals, as to let the shining eyes of Aphrodite meet the bronze-eyed gaze of Athena.[27] A negative reading can show how Adonis, by virtue of his behavior on the hunt, finds himself excluded from the masculine world instead of seeing himself integrated within it as befits an adolescent of his age. But to insist on a shortcoming, this interpretation runs the risk of forgetting that the fair eyes of Aphrodite are not strangers to the conduct of Adonis the hunter.

Ovid's version is very explicit on this point.[28] Seduced by the young man's beauty, Aphrodite forgets the strand of Cythera, abandons the heavens, and dogs the steps of Adonis. To follow her lover in his race through forests and over mountains, Aphrodite transforms herself into a mistress of the hunt. With her dress tucked up to her knees in the manner of Artemis, she rouses the dogs and pursues the animals—not without discrimination. Her quarry are the animals "that can be taken without danger: hares quick to flee, head down; stags with their tall antlers; or even does. She keeps her distance from boars, frightened of their force; she avoids rapacious wolves, bears armed with claws, and lions who glut themselves with the blood of cattle."[29] Upon introducing herself into the world of the hunt, Adonis's mistress straightway effects a division of the animals into two groups: on one side, hares and deer, quarry that flee before the hunter's onset; on the other, bears and wolves, lions and boars, fierce creatures whose aggressiveness provokes flight. Aphrodite does not trace these limits on the hunter's sphere for herself alone. She assigns them to her lover through a myth she tells him in order to justify the treatment reserved for the most savage of the animals, in particular the lion and the boar. If these fierce creatures are excluded from the circle of game to which Adonis finds himself confined, it is because Aphrodite avows a merciless hatred for them that, in return, the violence of these savage beasts does not cease to swell. At the origin of the division stands a young woman named Atalanta who dedicates herself to the hunt because she abhors both Aphrodite and marriage—an abhorrence that will cost her metamorphosis into a lion, one of the animals whose odious race poses a heavy threat to the happiness of lovers.

So between Adonis the seducer and the huntress Atalanta a confrontation becomes inevitable on two levels, one of hunting and the other of amorous desire. Aphrodite in person provokes and demands it. Moreover, the meeting between Atalanta and Adonis is neither surprising nor without precedent. An Etruscan mirror of the fourth century B.C. gives a version of it that stands beneath the sign of the hunt (fig. 1).[30] In the center, at the top of the scene, the head of a boar; disposed on each side, two clearly separated couples; on the right, Atalanta and Meleagros: he is standing, leaning a shoulder gently on his lance; she, naked like her companion, is seated, leaning an elbow on her right knee; turning herself from what surrounds her, she lifts her gaze into the distance. The other couple is likewise heterosexual, but this time it is the male personage who finds himself seated in symmetry to Atalanta. In his left hand he grips a lance, while his right draws toward him the stooping woman whose arm already embraces his shoulders. The inscription names her Turan.[31] It is the Etruscan Aphrodite tightly entwined with her lover Adonis. Between the two couples and on the axis of the monstrous animal a

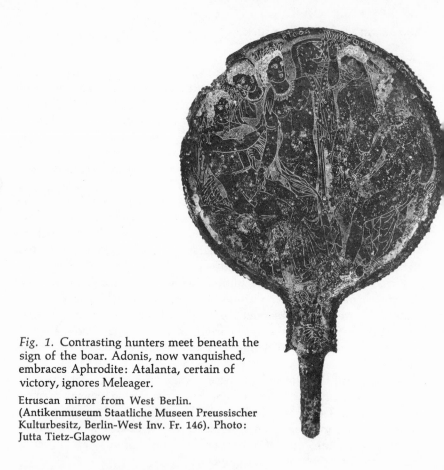

Fig. 1. Contrasting hunters meet beneath the sign of the boar. Adonis, now vanquished, embraces Aphrodite: Atalanta, certain of victory, ignores Meleager.

Etruscan mirror from West Berlin. (Antikenmuseum Staatliche Museen Preussischer Kulturbesitz, Berlin-West Inv. Fr. 146). Photo: Jutta Tietz-Glagow

Fig. 2. Aphrodite, with Hermes attending, gazes after Persuasion, her envoy to a hunter with panther in tow.
Pyxis of the Painter of the Würzburg Amymone. (H 5333, Martin von Wagner Museum der Universität Würzburg)

Fig. 3. Lover with panther. Amid loving couples of both sexes or the same sex, a beloved receives the gift his lover offers.

Kotylos of Amasis (Musée du Louvre A 479). Photo: Chuzeville

fifth personage occupies the center of the scene, a winged woman with a hammer in one hand and, in the other, a nail she is preparing to drive at the height of the boar's head. This is Atropos, the Fate, fixing an irrevocable destiny not only for Adonis and Meleagros but also for the two contrasted couples. Atalanta, seconded by Meleagros, triumphs over the savage Calydonian boar that mobilizes the flower of the youth. Inversely, Adonis dies a pitiful death, conquered by one of those creatures of the woodlands whose hatred Aphrodite ordered him to fear, whose violence she demanded he flee. The dominant opposition is not between Adonis and Meleagros.[32] The latter has a minor role, wedded as he is to the tragic end that Artemis's displeasure has reserved for him by turning him against his maternal uncles when they refuse to allow a woman to win the boar's hide and rob them of a trophy symbolic of masculine virtues and power. In contrast to the hunter whom a "shameless passion" chains to the body of his mistress,[33] Atalanta is presented not simply as the huntress without equal among men but also, in the distant, fleeting link she forms with Meleagros, as a woman indifferent alike to the ferocious beasts who inhabit the forest and to her companion's desires.

Ovid's version of the adventures of Adonis is interesting not solely for the series of differences it develops between a hunter diverted into a byway of the hunt and a huntress who flees marriage. In addition, it orients this development in a direction that maximizes the opposition between the two personages, since Aphrodite's narrative, as it forewarns Adonis of his death, also imputes responsibility for it in advance to Atalanta, but discreetly, through the implicit homology in the Greek bestiary between lion and boar.[34] Paradoxically, Ovid's narrative is not centered on the exploits of Atalanta prowling through the forest, defying savage beasts. Instead, racing has replaced hunting. Atalanta is a young person very skilled at footracing. She is even so gifted that she surpasses all male champions in the event. Perhaps that is why one day Atalanta goes to ask the oracle if she must take a husband. The response is formal: you do not need one, and even, flee him! To this advice a threat is added: nevertheless you will not escape, and, without ceasing to live, you will cease to be yourself.[35] Terrified, Atalanta runs away. She withdraws into the dim forest, only emerging to submit impatient suitors to the trial of speed whose stake is her virgin body or the man's head.

From antiquity mythographers wished to distinguish two Atalantas: one, born in Arcadia, hunts with bow and arrow; the other, a Boeotian, excels in running.[36] It is undeniable that different versions of the same myth can vary, here as elsewhere, in their geographical components. But running and hunting, where a woman is concerned, are not activities so clearly contrasted that they alone justify a radical division. Besides, Ovid's narrative does not separate them: once she decides to flee marriage, the forest becomes the domain of the woman who is too swift of foot. In a previous story, the *Metamorphoses* have told of the part taken by Atalanta in the expedition against the Calydonian boar.[37] A girl who runs faster than her male competitors must of necessity appear in her own eyes and those of others as an ambiguous being with respect to gender. When he sees Atalanta join the hunters tracking the boar, Meleagros does not know whether the face before him is a virgin's on a young man's body or a young man's on a virgin's body.[38] Doubtless the footrace is a less clearly masculine activity than the pursuit of wild animals. The little girls sequestered in Artemis's sanctuary at Brauron,[39] like the college of sixteen wives at Olympia and the young maids of Sparta who compete at the festival of Helen,[40] participate in trials of speed.[41] But in none of these contests do the women compete with masculine adversaries for whom swift arms and legs are the marks of a warrior. On the contrary, the races in which Atalanta triumphs unfold exclusively on male ground, where her superiority first leads her to query her sexual identity, then permits her to eliminate suitors by revealing that despite appearances she does not belong to the familiar world of women.

In the case of Atalanta, the homology between racing and hunting is even clearer since in both activities she bears arms and causes blood to be shed. In reality, the race Atalanta imposes on her suitors is only a continuation of the hunt by the same means and with the same weapons. In the version preserved by the *Library* of Apollodorus, her refusal of the world of women marks Atalanta's destiny from the start. Her father, who wanted a male child, exposes her. A she-bear suckles and cares for her until some hunters discover her and take upon themselves her education.[42] Before triumphing over the boar and taking the form of a lion, Atalanta is a baby bear—not after the fashion of Athenian girls, who "do the bear" before marriage in honor of Artemis of Mounikhia or Brauron to purify themselves, in the words of an ancient exegete, of any trace of savagery.[43] On the contrary, in sucking the milk of a she-bear, Atalanta is literally introduced into the world of savage beasts and, simultaneously, shanghaied from any vocation for marriage, which would certainly not recall to her the companionship of men exclusively impassioned by the hunt in the depths of the forest.

"Having become a fulfilled (*teleía*) woman, she wished to remain a virgin (*parthénos*), and hunting in the lonely forests, she never abandoned her arms (*kathōplisménē dietélei*)."[44] The opposition indi-

cated even in Apollodorus's narrative between physiological fulfillment of puberty and the goal chosen by Atalanta, armed virginity, is
broadly exploited in the version given by Theognis of Megara.[45]
Atalanta is ripe for marriage (*hōraíē*), but she refuses it and escapes
a wedding (*gámos*)." Having girded herself [or, equipped herself with
arms] (*zōsaménē*), she fulfilled endless [or, useless] exploits (*atélesta
télei*). Abandoning her father's home, blond Atalanta went away over
the high mountaintops to escape desirable union and to flee the gifts
of golden Aphrodite. But despite her refusal, she experienced marriage [or, understood its fulfillment] (*télos d'égnō*)."[46] The season of
marriage opens with sexual maturity (*hōraîos*). Once growth permits,
that is, once the woman is fulfilled (*teleía*),[47] the conjugal state is
obligatory as the natural fulfillment of the female being, just as a
tree, once it has reached maturity, cannot but bear fruit. On the
contrary, Atalanta refuses to fulfill herself in marriage, and in this
case, the homology between *gámos* and *télos* is justified by the syntactic parallelism that frames Theognis's narrative of the adventures
of Atalanta.[48] In her father's home, virginity is untenable. Marriage
will sooner or later do her in. The only possible way out is to short-
circuit the fulfillment that threatens her by forestalling it with another,
the one that opens on a strange terrain where conjugal relations have
been abolished, where a woman's image, once she penetrates this
space, is blurred and no one knows whether she is girl or boy.

Ambiguity colors these actions from the start. The expression "gird
oneself" tends in two directions. The girdle is the last article of
clothing put on. To knot or tighten it completes a woman's toilette
or ends the arming of a warrior. When the gods convoked by Zeus
crowd around Pandora, the snare he has devised to deceive Prometheus
and mankind, Athena supervises the maiden's adornment and ties
her girdle.[49] But at Thebes, in memory of a counterattack led by
Amphitryon against the warriors of Euboia, a statue was erected of
Athena "of the girdle" (*Zōstēría*) in the very spot where the Theban
king had been fitted out and had donned his armor.[50] Between the
adornment of a woman and the arming of a warrior, Athena, virgin
and warrior both, assures the more effective mediation since she
prefers girding and knotting, gladly abandoning the privilege of the
untied virginal girdle to Artemis;[51] for this piece of clothing is also
the one which the husband is the first to untie and which the young
woman must consecrate the first time she has union with a man.[52]
Just so Atalanta has no intention of untying hers, and she only girds
herself to keep and defend her virginity.

Upon entering the domain of the hunt and putting on the arms
she will thenceforth never cease to wear, Atalanta ties around her

waist "the girdle of Ares," the talisman that assures the Amazons, through their queen Hippolyta, preeminence in the practice of war.[53] Instead of the fulfillment of marriage (*télos* . . . *gámoio*) she chooses to fulfill (*teleîn*) exploits whose essential virtue is to be deprived of conclusion and limit. They are *atélesta* in two senses: without end, since they must never cease,[54] but also fruitless, since they are vain and useless. Atalanta's hunt is interminable just as the race to flee marriage has no finish. In the same way all acts of prowess in the hunt are useless once they are no longer directed toward an ultimate return and integration within civilized society.

Like the Amazons, who are carnivorous virgins,[55] the huntress Atalanta fully assumes the traditional double sense of the epithet *antiáneira*: similar to the male and hostile to man.[56] A woman who is equal to a man can only be his enemy, in running as in hunting. But precisely in the trial of speed Atalanta reveals a ferocity that the pursuit of wild animals does not satiate but instead goads on to extremes. In the version of Apollodorus, once Atalanta rediscovers her father, he immediately urges her to answer the suitors' hopes. Atalanta takes command, imposes the test, dictates the conditions. The man has a few lengths' headstart,[57] but it is only the self-imposed handicap of a champion who wishes to temper her already too clear superiority. Between the two adversaries there must be some distance, for without it there would be no chase. In reality, Atalanta starts her game at arm's length. The man is naked; she is armed for the hunt, with lance, javelin, or dagger.[58] On the side of a glass of Roman manufacture kept in the castle of Vincigliata near Florence, Hippomenes hurtles at top speed toward the finish with his head turned back at his pursuer. She, in turn, springs forward with an unsheathed sword in her right hand.[59] Thus the race to which the suitors are invited is only a kind of hunt in which they are obliged to play the role of quarry, of the harried beast that only owes its safety to the swiftness of its feet.

The ceremonial hunt in which Atalanta indulges is not reducible to perverse passion for an activity that so violently denies the conjugal model. The test of speed is not unrelated to marriage. Ulysses wins Penelope at Sparta after winning a footrace that serves as a casting vote in the choice among numerous suitors.[60] In the Libyan tradition, Antaios, King of Isara, posts his nubile daughter at the finish of the racecourse and declares to the assembled suitors that the man who first touches her veil wins her hand. At Argos, the forty-eight daughters of Danaos are married without delay thanks to the wisdom of a father who decides that each of the candidates who has come to compete will obtain in order of finish the hand of one of his daughters.[61] In a trial of this kind, the woman does not run. She remains at the end of the track, in front of the suitors who compete for her in agility and vigor. Deliberately, Atalanta inverts this scenario. Instead of offering at the goal of the race her desire-awakening body to the fastest, she is the woman-in-arms chasing before her naked men startled into mad flight. And Atalanta requires of her uncertain victor only that he be the fastest to flee before her. So the subversion of marriage by hunting operates on a terrain where Atalanta, carnivore and virgin, by inverting the accepted roles of male and female, can flatter herself for not hunting otherwise than his mistress instantly recommends to Adonis: choosing the quarry that can be captured without danger, tracking only timid hares and timorous does, all beasts quick to flee before the hunter.

Such is the huntress whose adventures Adonis hears told by Aphrodite as she leans on her lover's bosom, languishing from the exertion of hunting hares—a tale, says Ovid, that was often interrupted by embraces.[62] While for Atalanta hunting is the chosen means of denying amorous desire and refusing the gifts of Aphrodite by forcing the space reserved for marriage to become nothing more than the hunter's domain, for Adonis and for the mistress who pants the rules of the game into his ear hunting is an excursion into the countryside that alternates between the amusement of chasing a panicked hare and the pleasure of stretching out on the grass in the shade of a big tree. As much as Atalanta in devoting herself to the hunt wishes and makes herself more manly than a man, by so much Adonis in the same activity finds the chance to prove himself effeminate and voluptuous, a hunter so seductive that Aphrodite abandons the sky to accompany him and forbids him to attend to any victims but those that oppose his ardor with as little resistance as a lover hopes for from his mistress. The choice of certain animals accentuates further the erotic nature of the Adonisiac hunt, for deer as well as hares are among the denizens of the forest offered in Attic pottery as presents to the beloved. But here again, if comparison confirms the aphrodisiac symbolism of the prey reserved for the courage of Adonis, it also makes clear how the erotic relationship between Aphrodite and her lover belongs in the domain of the hunt on the masculine and homosexual model of a relationship between *erastés* and *erómenos*, between adult warrior and still beardless adolescent.

It follows from the logic of the split made by Aphrodite in the hunt that as effeminate a hunter as Adonis should become a marked victim for ferocious beasts, lions or boars. The narrative has only to decide the modalities and the occasion for an encounter. The three versions of the death of Adonis seem to arise in such a context. In the first,[63] the principal role is held by the Muses, daughters of Memory, who wish to avenge themselves for being forced by Aphrodite to couple with mortals and bear their progeny. Their weapon is song; they devise a hunting song so enchanting that Adonis, filled with pride—or blinded, in a secondary account—rushes forth to confront the monsters of the forest. The boar was at the rendezvous. To induce forgetfulness of Aphrodite's prudent counsel, the magic of a hunting song is necessary. And the plan devised by the daughters of Memory would be useless were not Adonis a lover ignorant of the virile hunt dramatized by Ovid's narrative and the Etruscan mirror.

In the second version,[64] Artemis takes the initiative. She seeks vengeance from Aphrodite for causing the death of another hunter, Hippolytos, who, unlike Adonis, had chosen Artemis's forest to preserve himself from marriage and to refuse sexuality with as much passion as Atalanta. The pleasure trip Aphrodite organizes for herself and her lover could only exacerbate the anger that the mistress of the hunt was already nurturing. As for the third version,[65] it combines death from the charge of the enraged beast with the special punishment of a seducer. Ares, Adonis's unhappy rival, takes the form of the boar, putting its warlike capabilities in the service of a rejected lover's rancor.

A recently published ceramic source specifies further the privileged relationship that binds hunting and seduction in the myth of Adonis. In 1972, Erika Simon brought to light the circular imagery of a small box (see fig. 2), a pyxis that can be dated to 380 B.C. and attributed to the Painter of the Würzburg Amymone.[66] The frieze develops between two seated figures of opposite sex in a vegetal decor that situates the scene in nature, and more precisely, in a place reserved for hunting, as the two lances in the left hand of the seated young male indicate. Around and about the two principal figures, four personages are disposed in twos: two young, winged cupids, the first of which, holding an animal on a leash, turns toward the hunter whom a standing woman addresses with gestures betraying speech. The other cupid, kneeling, is the pendant of a masculine personage with *pétasos* and caduceus who stands beside the seated woman. Simon's interpretation has permitted identification of all the personages. Thanks to his caduceus and headgear, Hermes is doubtless the least enigmatic. Standing with an elbow leaning on his slightly raised knee and his right hand on his hip, the god fixes his gaze by a turn of the head at the seated female figure. Two traits permit the identification of Aphrodite at Hermes' side: the cupid kneeling at her feet and the gesture of her right hand, which is raised somewhat over her shoulder as though holding back a veil. The posture is characteristic of Aphrodite "of the Gardens."[67] The goddess's gaze is fixed upon both the young woman and the handsome hunter who, in turn, turns his head toward the two women to attend to his interlocutor. An *oinokhoë* from the Hermitage,[68] dated at the end of the fifth century B.C., leaves no doubt as to the sense of the scene. The woman so persuasively addressing the seated young man can only be Aphrodite's faithful companion, Persuasion (*Peithô*) herself, the messenger "who has never been refused."[69] On the relief vase in the Hermitage the same divinity occupies the medial position between two personages whose names are given by inscriptions: Aphrodite on the one hand and, on the other, seated before her with an Eros on his knees, Adonis. To identify with greater precision the hunter to whom Aphrodite is wafting her "honey-worded spells,"[70] the editor of the pyxis could not neglect the feline held on a leash by the cupid symmetrical to Aphrodite's acolyte. It happens that this carnivore, apparently trained for the hunt and resembling a feline of the genus *Panthera*, presents the zoological features of the cheetah.[71] In ancient Cyrenaica, which inherited its use from the Berbers, and also in Egypt since the eighteenth and nineteenth dynasties,[72] the cheetah was used as a dog both because it is easily domesticated and because of its remarkable running capabilities, which permit it to attain speeds of sixty miles per hour.

The hunter whom Aphrodite lusts after is therefore some oriental prince: Paris-Alexander, Anchises, or Adonis.[73] Since the first two ordinarily manifest their oriental origin by the dress and coiffure lent them by the iconographic tradition and since, in addition, both are herdsmen and not hunters, it is Adonis's candidacy that seems the most serious after all. It remains to interpret the meeting of Aphrodite and Adonis in a forest setting and to explain how the amorous embassy of Persuasion can be reconciled with the hunting project apparently specified by the feline's impatience to depart for the chase. The interpretation Erika Simon suggests centers the action around a single personage.[74] Adonis placed at the crossroads between twin contending cupids must decide who will win, the hunter's passion or the happiness of love offered by Aphrodite. Will Adonis yield to the call of the wild and the appeals of the cheetah, or, on the contrary, will he heed the persuasive voice of the go-between who has come to offer him the pleasures of marriage and the promise of union with Aphrodite?

The iconographic tradition of twin cupids and enemy brothers depicts *Eros* and *Anteros* at odds. At times one is blond and the other has black hair, but almost always they fly at each other or wrestle together, sometimes alone, sometimes beneath Aphrodite's eyes or before an audience of goddesses.[75] On the pyxis of Würzburg, however,[76] the two cupids ignore each other and are preoccupied with their function as acolytes to the two major personages, whose turned heads accentuate their symmetry. And there is no confrontation between Aphrodite and Adonis. As on the relief vase at the Hermitage, the persuasive speech of *Peithó* follows the path of a reciprocal desire that causes the two lovers to gaze at each other amid the forest. Aphrodite's look is relayed by the voice of Persuasion, whose triumph is foretold by the attention of the hunter turning his head toward the two women. Posted near Aphrodite as an attentive spectator, Hermes is not there simply by virtue of his time-honored complicity with Aphrodite and her companion. His presence is justified by an equal competence in the two domains that cross before his eyes as he contemplates Aphrodite's love for a young hunter in a wild place far from roads and homes, far from tilled fields. The Adonis of the Painter of the Würzburg Amymone has naught to hesitate between love and hunt; he has already made his choice, and the proof lies in the very cat his comrade Eros holds on a leash.

In reality the iconography of myths cannot be satisfied with bare zoological facts. It is still necessary to search for the place occupied by the cheetah in the bestiary of the Greeks, in that mixture of encyclopedic knowledge and symbolic values attributed to different kinds of animals. Besides, in figured representation, and speaking only of vases, it is not always easy to distinguish a cheetah from a panther. On a Vulci cup that belongs to the British Museum, an elegant ephebe carrying sponge and strigil at the end of a long stick moves forward, holding on a leash a superb spotted cat. For O. Keller, specialist in the animal history of antiquity, it is a cheetah.[77] For Erika Simon, it is instead a panther or a leopard.[78] The hesitation is not only permissible, it is indispensable. The Greeks for whom these images were destined were not endowed with the linguistic means to distinguish these different animal species from each other. Although they possessed specific terms for lion and for tiger, when it came to the members of the genus *Felis* (*Panthera* and *Acinonyx*), the Greeks used two words almost indifferently: *párdalis* and *pánthēr*.[79] When Aelianus in his treatise *On the Nature of Animals*[80] tells the adventures of a hunter who had tamed a *párdalis*, it is impossible to know if he meant a serval, a cheetah, a panther, or a leopard. Such semantic imprecision cannot be held of no account in the figured representation it corresponds to.

From Aristotle to the bestiaries of Byzantium, animals of the type *párdalis* and *pánthēr*—for simplicity we will call them "panthers"— present a certain number of clearly defined symbolic traits. As with other of the great wild creatures, the panther is not an animal that is hunted but an animal that hunts. The man who confronts it must be of like valor.[81] The panther sets itself apart from the rest of the animal world by its hunting habits, which ally it to the fox and to creatures of cunning. Like them it possesses the quality of prudence, *phrónēsis*, an intelligence that proceeds by subterfuge and knows to hide the goal it proposes to attain. Doubtless it cannot rival in this game the fox, who compliments the panther in Aesop's fable for his spotted dress, only to observe that the dapple (*poikilía*) of his hide is nothing compared to the dappled (*poikílos*) spirit that has won him, Renard, the surname "cunning."[82] Doubtless the panther likewise knows how to play dead to capture monkeys despite their agility and distrust of him.[83] But if the fox exhibits his prudence in the abrupt reversal of the dead being alive, the panther's deceit is more secret: it has recourse to the faculty of good odor. In fact, the panther is a perfumed beast, which also distinguishes it from all other animals. No creature naturally emits a good smell, writes Theophrastus,[84] except the panther. And an Aristotelian *Problem* poses the question, without actually answering it clearly, as to why the animals are all malodorous except for the panther.[85] In the town of Tarsus once there was even a prized perfume called "panther" (*pardalium*), but its formula had already been lost by the time of Pliny.[86]

The panther knows of his good odor and uses it to capture his victims. Aristotle explains it when giving some examples of prudent behavior in the animal world. "The panther, it is said, realizes that wild animals love to sniff its perfume; it hides to hunt them; they come right near it, and in this way it catches even deer."[87] Pliny insists on it: the panther is clever at concealing itself. For if the animals are all strangely attracted by its odor, its grim mien puts them to flight; which is why the panther is especially careful to hide his head as well as the rest of his body.[88] That invisible perfume is the trap in which victims come to ruin. By Aelianus's account, the panther is content to exhale a fragrant breath whereupon fawns, wild goats, and all the creatures of the forest approach as though bewitched, as though charmed by an *iunx*.[89] The technique is still more refined in the *Physiologus*,[90] which passed on to the Middle Ages the tradition of the *Dit de la panthère d'amors*:[91] once the panther has dined, it retires to its lair; three days later, it awakens, roars, and since its great voice is full of aromatic perfumes, immediately all the beasts, lured by the exquisite odor, run up to throw themselves into the half-open jaws of the wild creature who awaits them.

Such a fragrant creature cannot himself remain insensible to the perfumes that surround him. For this reason, it seems, panthers are frequently captured in the region of Pamphylia, which possesses numerous aromatic species. Attracted by the sweet emanations, panthers come over the mountains from Armenia, cross the Taurus, and proceed toward the gum of the styrax tree when the wind begins to blow and the trees exude their fragrant odors. Philostratus, to whom we owe this information, tells us in the Life of Apollonius of Tyana[92] how one day in Pamphylia one of the creatures was captured with a golden collar on his neck inscribed in Armenian: "King Arsacus to the god of Nysa." Because of its beauty and size, the panther was consecrated to Dionysos.[93] For a while it tolerated the caresses of its master, but springtime excited it, and it departed for the mountains. It was caught in the region of Pamphylia, lured there by the spices.

To secure one of these wild creatures who lust after good smells, men have only to turn against them their power to seduce and deceive. Oppian gives the key ingredient: the aroma of wine.[94] If a few bottles of it are sprinkled near a water-source, the panthers, roused by the smell, draw near and drink all there is. It remains only to take advantage of their drunkenness and capture them. If the trap is so effective, it is without a doubt because panthers also have a reputation for always being thirsty; given the dry constitution demanded by their good odor, they cannot but have a perpetual thirst.[95]

In his hunting technique, the panther combines deceit and seduction. The trap he lays for his victims is nothing other than his wild beast's body, whose perfume brings forgetfulness of the voracious death it conceals. This seduction by smell must have produced the intimate association of the panther with the image of the perfumed, full-bodied woman. For Aristophanes and his contemporaries a courtesan is in reality a "panther" (pórdalis).[96] But the word lacks the insulting and scornful connotation of the word kasalbás, which refers to skin and bedding as well. In the Lysistrata, after the men are forced to give way to the arguments of their spouses and submit to defeat, the male chorus leader renders irritated homage to the desire that female bodies arouse: "No beast more indomitable, no fire more ravenous, no panther so bold." Lysistrata's friends have just shown that with transparent blouses, little saffron tunics, and perfumes, they are capable of "roasting" and "broiling" their husbands.[97] As for Myrrhina, who outdoes herself by drenching her body in perfume and then letting her husband eat his heart out—does she not testify to a greater fierceness than the panther's toward the victims summoned by its odor? Besides, Socrates is the one who, scandalizing the prudish, devises the theory for this practice at his interview with Theodota, a very beautiful courtesan who was Alcibiades' mistress.

He explains to her that she submits to an activity of the same order as the hunt to find herself lovers. She uses beaters, dogs of various sorts—even nets are at her disposal. And all this equipment to track and entrap her prey is only her woman's body, beautiful and yearned for.[98]

Like the panther, the beautiful courtesan practices a kind of hunt that the Greeks call "the hunt of Aphrodite" (aphrodisía ágra).[99] Desire is its snare, and its victim is seized with love, just like partridges, who are so passionate that the female's piercing love call arouses in the male such violent desires to copulate that he even at times alights right on her head.[100]

The Adonis of Würzburg is not, then, like Herakles at the crossroads. His choice has already been made—a choice for the hunt, but a hunt after the fashion of the panther whom the cupid holds at his side. Answering the persuasive voice of Peithó is the symmetrical, bewitching call of the beast who knows no difference between hunting and seduction. It is as if in a single figured scene the painter had wished to prolong the myth Ovid's narrative provides, to show the links that join the lovers on both sides to seduction, and to inscribe in the hunter's domain familiar to Adonis an unprecedented homology between the desire aroused by the product of the myrhh tree and the fatal attraction irresistibly exercised by the perfumed breath of the panther.[101] No other animal, it seems, could better symbolize the aphrodisiac essence of the hunt reserved for Myrrha's son, a hunt as unmasculine and unmanly as that practiced by the courtesan rightly surnamed Panther.

The Wind Rose

The confrontation between Atalanta and Adonis develops not under the sign of marriage, which the former would reject and the latter obtain, but under the puissant and exclusive sign of sexual pleasure and Aphrodite, for which the handsome hunter evinces a propensity just as violent as their avowed abhorrence on the part of the virgin slayer of the monster of Calydon. Aphrodite herself guarantees this when she takes on the narrator's role and introduces the story of Atalanta to justify the limits assigned to Adonis's exploits and also to evoke in advance the inevitable failure of the division she has just finished making. The new presence of Adonis's panther invites further analysis of Atalanta's adventures, especially their culmination, as reported by Ovid, in the metamorphosis into a lion of the huntress on whom Aphrodite imposes, by deceit and on her own wild terrain, desire's constraint.[102]

Aphrodite answers the warlike violence that Atalanta trains on her suitors with the weapons of deception, with *feminine apátē*, with golden fruit. The suitor whom Aphrodite swells with her favors is Hippomenes to some, Melanion to others. Hesiod names the first of the two, and Ovid follows suit, but this Hippomenes is a neutral, colorless figure, in contrast to his rival as found in the version of Apollodorus.[103] In fact Melanion, the Black One,[104] is of the same species as Atalanta. He is a hunter who refuses to take a wife. On the François Vase, Melanion clings to Atalanta's side behind Meleagros and Peleus.[105] But for the old men of Aristophanes who take courage from his example, the Black One is a misogynistic boy who lives celibate in the mountains' solitude, hunting the hare with nets he has woven himself. "From hatred of women he never went home again, so much did he abhor them."[106] Meleagros is sociable and addresses the boar with a lance, while his rival flees marriage and the world to devote himself to a kind of lonely, devious, and dangerless hunt whose sole quarry is the hare. Plainly, then, the misogynist Melanion is Atalanta's double on the masculine side, a hunter from hatred of marriage.

For Aphrodite it's a question of two birds and one stone. Golden fruit in hand, the black hunter is the net into which Atalanta casts herself. The hare also beckons not only because Melanion captures it in his cunning nets without throwing a weapon but because it is one of Aphrodite's favorite animals and the present a lover offers his beloved. In the versions that have come down to us, Hippomenes and Melanion are merely figurants. The stratagem comes from Aphrodite, and she pulls all the strings. In the *Catalogue of Women* attributed to Hesiod and reconstituted from citations and papyrus fragments,[107] the story of Atalanta and Hippomenes centers around a gift that is first refused and then imposed on its recipient by devious means. Atalanta mistrusts sexual pleasure; "the gifts (*dôra*) of golden Aphrodite" are rejected.[108] But Atalanta yields to the persuasive voice that summons her to accept the golden fruit: "Receive these sparkling presents (*dôra*) from golden Aphrodite."[109] This time the gesture of giving accompanies the speech and serves to render it effective while at the same time setting its own plot in motion. The apples, which are thrown three times, hinder Atalanta from bridging the gap that separates her from her suitor. If she loses the race, it is because she stopped to pick up the fallen apples from the ground. In Apollodorus's version, inviting words are lacking. There remain only gestures, as if the apple were an irresistible trap:[110] Atalanta is compelled to follow the path of one fruit after the other. But in other texts, as in Ovid or Theocritus, effectiveness is concentrated in the apples themselves. Instead of causing Atalanta's defeat and thus forcing her to marry the winner, the golden fruits act directly, like a drug or an incantation. The reversal is complete: Atalanta is dispossessed of her superiority in running, and the contest takes place, if at all, only to restore it to its original form in which the fastest suitor carries off the wife he has thus won. For Ovid, Atalanta is defeated even before the race begins.[111] Having barely seen Hippomenes, "she loves without doubting that she loves."[112] Even so she has to chase after each of Aphrodite's apples. Ovid adds a moment of suspense: when the third apple rolls far off the track, Atalanta wonders if she should go and get it, but Aphrodite is there: "I forced her to."[113] For Theocritus, who compresses the action into two verses, Aphrodite no longer even needs to intervene. "Hippomenes, who wanted to marry the virgin, took the apples and ran the race. But Atalanta, in what madness (emánē), to what depth of love (es bathùn érōta) did she hurl herself at their sight."[114] We are beyond the word of persuasion. One look, and by the most vulnerable point of her being, her eyes struck by the apple's glare, Atalanta is filled with the madness of Eros.

The aphrodisiac symbolism of the golden apples is deployed between two extremes: on the one hand, the coercive violence of the gift that cannot be shunned, on the other, the courteous suitor's offer of a present that his intended seems freely able to accept or refuse. It is important to review the symbolic contexts of the apples within this spectrum, given the bonds of deception that unite Atalanta and Aphrodite at the crossroads of hunting and marriage.[115] There are two contrasting traditions as to their origin. In the one Ovid's narrative follows, Aphrodite herself picked them in the center of the island of Cyprus on her estate at Tamasos in the middle of which rustle the golden boughs of a resplendent tree.[116] According to the other, the fruit come from the Garden of the Hesperides.[117] But the sameness of the fruit cancels the geographical difference. For the only tree that Aphrodite nurtured at Cyprus is the pomegranate (Punica Granatum),[118] the fruit tree that a whole tradition associates with Hera's orchard and through it with the Garden inhabited by the daughters of Atlas. In her sanctuary at Argos, Hera in state holds a scepter in one hand and a pomegranate in the other,[119] a fruit from the tree that is said to grow only for her, the mistress of legitimate marriage.[120] In fact, at the time of her marriage to Zeus, who made her the unbending guardian of the conjugal bed, Hera received from Earth, who came with the other gods to give wedding gifts to the new bride, the golden fruit that thenceforth would shine in the garden of the gods beneath the protective gaze of the Hesperides, the Virgins of the ends of the earth.[121] Lest there be any misunderstanding, the fruit of the pomegranate tree, which pleases Aphrodite no less than Hera, is only another name for the "apple" in the fable. The Greek word for "apple" (*mēlon*) designates every kind of round fruit resembling an apple, and consequently, it is used not only for the fruit of the apple tree but for the pomegranate and the quince, which was known to the Greeks as the "Cydonian apple."

In the domain of marriage, where Aphrodite's power accommodates Hera's, sparkling round fruits—quince, pomegranates, and apples—are used in various ways in gestures and ritual practices. In one of his poems, Ibykos of Rhegium evokes happy love and the untouched garden of the Virgins where pomegranates and apples of Cydon ripen.[122] They are picked at wedding time, as is shown by the tablets of Lokroi: two young women are filling baskets placed at the foot of a tree whose branches are covered with round fruit. An attentive examination of different examples of the same scene has made it possible to recognize on the tree an alternation of pomegranates, quince, and plain apples.[123] These fruits are offered to the young couple and sometimes thrown at the wedding procession; for instance, the cart that bears Helen and Menelaos is covered with myrtle branches and apples of Cydon.[124] They know other uses as well. Freshly picked fruit are poured into the bride's garment, or a young woman accompanying the bride and groom holds a fruit between two fingers and presents it to them. At Athens, the ritual gesture was even sanctioned by the Solonic code, which enjoins the bride to munch an apple of Cydon before crossing the threshold of the bridal chamber.[125] Plutarch's exegesis may attribute to Solon the wisdom of thus assuring that the bride has a clean mouth and sweet-smelling breath,[126] but Persephone's experience in Hades is doubtless more revealing as to the symbolism of the fruit the bride eats. The gods have decided to return her daughter to Demeter. Hades then must assent, but before letting Persephone join her bereaved mother, he gives her a fruit to eat. It is a pomegranate seed.[127] Henceforth, Persephone will have to spend part of the year in the land of the dead, for she has become the wife of Hades. Persephone will tell her mother that her host did her violence by forcing her to eat a sugary sweet food, when actually the only compulsion inflicted upon her was that of the gift she received from Hades' hand without knowing it.[128] To be sure, there is deception in his way of offering the young woman her nuptial fruit: Hades glances round, he acts in stealth. But the essential point is that in the marriage ceremony the offering of a round fruit, be it quince or pomegranate, ritually consecrates the marital union. Persephone's misadventure provides proof of the gesture's effectiveness.

This effectiveness is apparent not only in the sphere of the marital contract presided over by Hera but also in the domain of desire, when Aphrodite intervenes in her own distinctive way. The medieval tradition calls it "requisition of love";[129] apple throwing, or in Greek, mēloboleîn, is a proverbial expression that to ancient exegetes signaled Aphrodite's mode of action: "render passionate, put into ecstasy, entice with the lure of sex" (eis aphrodísia deleázein) are their definitions.[130] The apple of love is a frivolous game when a girl tosses it at an open-mouthed boy.[131] But when in the sanctuary of Artemis the fair Ktesylla leans over to pick up the fruit thrown by Hermokhares, she can do but one thing: yield to the desire of this man who has loved her since the day he watched her dance round Apollo's altar. Furthermore, to precipitate her capture, the young man has scratched the following oath on the apple: "I swear by Artemis to marry Hermokhares of Athens."[132] For the Greeks, to read is to speak, and the bonds of the oath once spoken render the requisition of the apple still more binding. In Theocritus's version, the violence of love makes ruse useless. The mere sight of the apples overwhelms Atalanta. Whether poison or love charm, all is pure Aphrodite, but all is outside the domain of marriage, where desire takes on the appearance of a contract. For if there is fraud and deceit, it is there to catch Atalanta in the trap of marriage. Hesiod's narrative carefully traces the circle of marriage by means of the almost legal formula Hippomenes addresses to Atalanta: "Receive the shining gifts of golden Aphrodite." But behind the gesture that borrows the serene solemnity of a legitimate contract from the Hera of marriage there lurks the sudden madness of erotic desire. Under the deceptive guise of a relation presented as an accord and reciprocal gift between two parties, Aphrodite lets loose her natural violence, the violence that seizes, at her passing-by, "grey wolves, tawny-maned lions, bears, swift panthers" and makes them all pair off and couple in the deep shade of the valleys.[133]

To a certain extent, Atalanta aims her hatred of marriage so specifically at Aphrodite that it inevitably entails the erotic madness that propels the huntress into marriage and particularly into sex. But the wedding of Atalanta and Hippomenes is a failure. An oversight disturbs the progress of the ceremony. Normally, several sacrifices are required in the marital rite: in honor of Zeus and Hera, Artemis, the Charites, Aphrodite, and Persuasion.[134] The husband commits an error. He forgets to thank his protectress. Aphrodite is expunged from the ritual, and at the wedding she has prepared it is she who

is forgotten.[135] Her vengeance is immediate. Moreover, her means
are the same as for Atalanta's deception: immoderate desire. This
time the goad is plunged into Hippomenes. The versions equivocate
on the name of the sanctuary. Is it that of Cybele, mother of the
gods?[136] The temple of Zeus Conqueror on Parnassos?[137] In any case
it is a holy place cloaked in forests from which Atalanta and her
husband suddenly appear. One of the versions specifies that Aphrodite
punishes them in the course of a hunt, as if Atalanta had found in
the husband forced upon her a companion in her tracking over mountains and through forests.[138] The two have just stopped to rest.
Hippomenes feels a sudden, irresistible urge to make love. The sanctuary at hand seems inviting. For these newlyweds there will be no
union in the bridal chamber, but rather a wild copulation beneath
"the indignant gaze of the ancient wooden statues." Atalanta and
her accomplice behave like beasts. On this point Greek opinion is
unanimous and severe: it is forbidden to make love in a sanctuary
or even to enter a holy place without washing oneself after sexual
intercourse. Only animals, says Herodotus,[139] from cattle to birds, can
be seen coupling in the sacred enclosures. Plainly, Aphrodite wants
to do these lovers in. They run from the forest that could hide them
and stray from the clearings where Adonis and his mistress dally in
love. The anger of the gods bursts down upon their sacrilege, and
they are transformed into ferocious beasts: "their necks are covered
with tawny manes; their fingers twist into claws; from their shoulders
they sprout paws; the whole weight of their bodies presses on their
chests; they grow tails to sweep the sand; their gaze conveys anger;
instead of words, they bellow; instead of a bridal chamber, they
prowl the forests." Atalanta and her companion have become lions,
a terror to all and especially to young boys like Adonis. At the end
of her story, Aphrodite leaves him with the suggestion that he avoid
a race that is hateful to her.[140]

At the conclusion of her adventures, Atalanta finds herself confirmed in her vocation for the hunt and, as a consequence, in her
refusal of marriage. In taking on the appearance of the wild creature
that is the only hunter among animals neither men nor beasts take
as prey, she becomes queen of the hunter's world where hatred of
Aphrodite drove her and from which the same divinity's wiles failed
to expel her. In the Adonisiac context in which Atalanta's adventures
transpire, the place to stop is where Aphrodite and her machinations
abort. Indeed, certain versions add to the animal transformation
scene a behavioral trait that directly throws into question the relation
between Aphrodite and Atalanta.

Ovid's tale, which is keyed to Adonis's pathetic end, insists on the fearful boldness of lions, which attack head-on like boars and tear to pieces friendly hunters who forget the wisdom of the hare. But other Latin mythographers explain the choice of this beast for different reasons than its valor in the hunt. Lions, they say, are animals the gods do not permit to copulate or which, since the day unseemly desire misled Atalanta and her spouse, are condemned never again to make love.[141] So Aphrodite's enemies are punished precisely where they sinned. They indulged in copulation in defiance of the most sacred practice: henceforth they will be deprived of sexual pleasure. In the light of this logic, no conclusion seems more satisfying. But the paradox is that Aphrodite's vengeance tends to reinforce the refusal of marriage Atalanta asserted from the beginning. At the conclusion of her tricky maneuvers, the deity of desire can only express her anger by refusing her enemy the very "gifts" the latter had never ceased to reject.

The interest of these versions lies in their semantic renewal of the animal metamorphosis as a function of Atalanta's original hostility to a marital scheme whose emblem is Aphrodite qua goddess of desire and sexual union. A would-be wife, the huntress becomes not just one of those ferocious creatures who precipitate the loss of Adonis; she transforms herself into a great, sexless carnivore, a frigid lion, a beast disabused of all sexual activity who therefore detests Aphrodite as she abhors him. In Ovid's version, as in all those that retail the triumph of the hunt, it seems that Aphrodite submits to the path of failure, for it is really she who always and everywhere pulls the strings. No other marriage divinity comes to remind her of her limits or claim title to obvious privileges. Yet the affair also concerns both Artemis and Hera, since Atalanta, once she enters upon marriage, cannot fail to traverse the provinces of the one and the other. Doubtless the shadow of Hera is present when the suitor inspired by Aphrodite pronounces the contractual formula that introduces and at the same time screens the requisition of love. But all happens as though Aphrodite, abandoned to her own devices and driven by grudge, holding to her visage the mask of keeper of the keys to marriage, has been swept away by the violence of her power. She only succeeds in proving her own incapacity to rule the territory of marriage without assistance. One cannot attribute to Aphrodite's clumsiness the erotic madness that seizes Atalanta at the sight of the golden apples. Nor can one blame the vengeance of a third party for the ritual oversight, which Aphrodite pays for in a marriage where her sole mistake is the desire to rule alone. To answer Atalanta's passion for the hunt Aphrodite unleashes the erotic drive whose violence marks the limits and blares out the truth of the marital

trick. No doubt the conclusion of the story is not imposed by its pro-
tagonists alone. Elsewhere, in the versions tragedy exploits, Atalanta
does not end her career in the depths of the forest. She "loves men,"
or is a woman of the world, or even a courtesan versed in bizarre
pleasures who pushes Meleagros to divorce and makes Melanion her
slave.[142] There is no lion magnificently indifferent to the pleasures of
love except in the logic of the contrast between Atalanta and Adonis,
a contrast pushed to its extreme in the common domain bounded by
the intersection between hunting and Aphrodite's power. To a certain
extent it is indeed the perfumed panther who calls up in turn a wild
beast scornful of the rapture of love.

But the two animals, though equally magnificent, are not placed
on the same plane. They differ first of all in respect to their privileged
relationship to Aphrodite and her values. In fact, if to Greek tradition
belief in the virtues of the panther's fragrance is one of the explicit
givens of the bestiary, the image of an obstinately chaste lion is
decidedly less familiar. It is even very bizarre. For the information
from Pliny's *Natural History*, where since Frazer some have tried to
discover the explanation for such singular behavior, insists instead on
the sexual activity of lionesses who, drawn by thirst to the edge of a
river, mingle with leopards and other savage beasts in heat, thus
provoking the anger of the lion. He at once recognizes the telltale
scent of the adulterer and sets out full tilt to punish his female; as
a result, says Pliny, the lioness takes care to cleanse herself of her
error in the river or else to follow the male at a respectful distance.[143]
There is reason, then, to believe that sexual deprivation is inflicted
on this animal species only in the myth featuring the metamorphosis
of Atalanta and her shadowy husband into lions.[144] It becomes a
hapax whose uniqueness should then be integrated into the analysis.
Yet if in the lion's case underlying representations of his frigidity are
less known to us than the coherent whole that is based on the
panther's perfume, perhaps one can indicate among the lion's traditions the point of departure from which the doctrine of its hostility
to Aphrodite could have been established.

The winged serpents of Arabia, who invade Egypt at the coming
of spring, are a fierce and harmful species. They would have long
since filled the earth had not their own fierceness contrived an ob-
stacle. For when these creatures copulate and the male is in the midst
of ejecting his seed, the female snatches him by the throat and does
not let go until she has devoured him. But the young, in their turn,
while they are still in their mother's belly, devour her, since by
gnawing at her entrails they clear a way out for themselves. The
species' endocannibalism prohibits their rapid increase. For Herodotus,

who reports such stories from the land of the Arab, this one is decisive proof that divine providence wished the fierce and harmful species to be the least fertile, just as it made prolific those that are timid in nature and good to eat so as to prevent them from being eaten to extinction.[145] The theory of Herodotus proceeds to persuade by two examples: the lion and the hare.

No animal is more prolific than the timid one with hairy paws. Everyone hunts the hare: savage beasts, birds, men; which is why "alone among animals, the female conceives though she is pregnant; some young are covered with hair in their mother's belly, others have none, others still are being formed in the womb at the moment when yet others are being conceived." Yet the lion, a strong, dauntless animal, is on the contrary one of the least prolific. The lioness whelps but once in her lifetime, and she has only one cub; and, Herodotus adds, "when she does whelp, she expels the womb at the same time as its fruit." "Here is the reason for it: when the lion-cub begins to move inside its mother's body, since it has by far the sharpest claws of all animals, it lacerates the womb, and the more it grows, the deeper it cuts with its claws; when the lioness is ready to whelp, nothing remains intact."[146] Nonsense, says Aristotle, who has good reason to see a fable, *mûthos*, in all this. And yet as such the opposition of the hare and the lion is not gratuitous even if the zoologist can remind us that "the lioness usually has two cubs with a maximum of six."[147]

The hare, by virtue of its miraculous fertility, belongs to Aphrodite.[148] An erotic painting described by Philostratus makes the reasons clear. In the middle of an orchard a group of cupids is amusing itself encircling a hare busy tasting fruit that has fallen on the grass. Bows are against the rules, and the object of the hunt is to capture alive the animal who is called "Aphrodite's favorite victim." Her preference is motivated by the hare's amorous activity, for the female is always pregnant and the male never stops spreading his seed. This animal has a persuasive quality that pederasts themselves recognize.[149] The figured scenes on vases confirm Philostratus's information. Desire, *Hímeros*, flies off over the sea escorted by handsome young men with long wings, one of them holding a hare by the ears.[150] Likewise the Painter of Meidias draws an interlaced couple, Aphrodite and Adonis, and beneath them a jumping hare that an Eros holds tightly in his bare hands.[151]

Herodotus's fabulous tale leaves hidden the explicit relationship of both animal species to Aphrodite herself. But from the tradition attesting the aphrodisiac character of the hare one can, if only negatively and within the limits of the opposition Herodotus exploits,

deduce if not a calling for chastity at least a certain reserve on the
lion's part with respect to the "deeds of Aphrodite." Moreover,
Aristotle himself is a witness, whatever he may say elsewhere: sterility threatens the lioness "who at first whelps five or six young; the
next year, she has four, then three, and so on to one. Then she has
none at all, as if the seed disappears with age."[152] In the mythical
space opened by the concurrent and contrasting hunting exploits of
Atalanta and Adonis, the lion and the hare combine with the panther
to define a bestiary where reference to Aphrodite is the dominant
principle. Between the aphrodisiac panther and the lion who lacks
desire, the hare plays a complex role in his relations with the divinities of sexuality. His amorous temperament qualifies him as an
effective gift between male lovers for whom intrigue cannot be
divorced from hunting. In addition, his timidity and fearful nature
predispose his serving as an emblem for the shy object of lust.
Finally, his prolific nature assures him a privileged place in the domain
of fertility ruled within marriage, at least in part, by the Aphrodite of
Persuasion and the bond of love. None of the surviving versions of
the conflicts in hunting and love between Atalanta and Adonis exploits the potential in the series of symbolic values for the hare, not
even those that suddenly cut short the sequence of the golden fruit
devised by Aphrodite's cunning. The Ovidian narrative, which has
imposed the framework of our analysis, is built entirely on the statement of extreme terms whose opposition is made more accessible by
the absence of mediators and operators that would favor displace-
ments or introduce zones of opacity within the story.

The metamorphoses in Ovid's version of the parallel destinies of
Atalanta and Adonis obey a certain logic, one of whose elements
further reinforces the rigor of the antithesis between the hunting
lover and the huntress hated by Aphrodite. It concerns, to begin with,
another aspect of the asymmetry between lion and panther. In fact,
though Atalanta takes on for eternity the form of a lion, Adonis does
not suffer the animal metamorphosis that would fix him at a point
where the hunt is seduction pure and simple. The panther disappears
once the fierce creature rises up whose quarry becomes the amorous
hunter; it leaves without a trace even of its perfume. Ovid's version
insists upon it, in a context in which Aphrodite's disappointment is
proportional to the ultimate treatment reserved for her lover which
is, in turn, compared to that suffered by the no less unfortunate
Mintha, Hades' concubine and the rival of Persephone. The teeth of
the boar lacerate his skin, the earth is spattered with blood, and
Aphrodite, in mourning, makes two promises. Each year the scene
of his death will recall her lamentations; the blood of Adonis will

be changed into a flower.[153] The metamorphosis takes place on the vegetal plane and belongs to the botanical scheme already constructed of spices, lettuce, and cereals. It functions in open competition with the transformation of Mintha into fragrant mint[154] and in implicit parallelism to the metamorphosis of Leucothoe into a stalk of incense. A cruel father had separated her from Sun, her distant lover, by burying her in a ditch, and in this way the couple is reunited.[155] Aphrodite's gestures are the same as Sun's: she scatters fragrant nectar.[156] And from the red blood of Adonis, which bubbles and seethes in the heat of the divine substance, there springs up, an hour later, a flower of the same color: "it resembles the pomegranate flower, which hides its seeds in a thin shell, but it gives no lasting pleasure since, weakly fastened and much too light, it falls to the ground, loosened by the wind that gives it its name."[157] This is the anemone, whose Greek name *anemónē*, suggests the word for wind, *ánemos*.[158]

A plant with a velvety thin stalk at whose peak opens a flower like the poppy's, with black or dark blue heads, the anemone is,[159] despite Dioscorides' warning, sometimes confused with the thorny or field poppy.[160] In the spring flowering season it is closest to the rose, the last to blossom and the first to wither.[161] But the two flowers have more in common than their brief efflorescence. Besides Ovid's version, other, not exclusively narrative traditions combine the rose and the anemone at Adonis' death. Thus roses are born from the blood of Aphrodite, who is scratched by thorns during her travels, while the white anemone is stained crimson by the blood of Adonis.[162] Elsewhere, in the *Lament for Adonis* attributed to Bion, it is the hunter's blood that sprouts the rose and the anemone that flowers where Aphrodite sheds her tears.[163] In this last version, the affinities of the rose and the anemone permit them to exchange roles, but at the cost of a choice in favor of perfume even unto death. For if the proverb says one must not compare the bramble or the anemone to the rose,[164] the reason is that of these two equally transient vernal flowers, one is fragrant, the other is not. The rose (*rhódos*) owes its name to the flood of scent (*rheûma tês odôdes*) it releases.[165] Its perfume is so violent that it makes vultures, who hunger for the scent of rotting corpses, die.[166] The anemone, on the other hand, is odorless, and the scholiast responsible for the exegesis of Nicander's version, in which this flower springs from the blood of Adonis, defines the anemone as a rose without scent (*ánodmon*).[167] In the context of Ovid's narrative alone, the anemone metamorphosis assigns to Aphrodite's lover the lowest position on the ladder of lovers transformed into plants: for Leucothoe, the incense tree, for Mintha, the fragrant herb, and for Adonis, the flower that is just an odorless rose.

Two other features accentuate the negative character of Adonis's metamorphosis. The red anemone is like the pomegranate flower, but the resemblance is only stated to mark the absence, in the flower born of blood, of the promise of fruit and also of the fertility that the fruit customarily offered to the bride on her wedding day hides under its thin shell.[168] Just like the cereals and seeds of Demeter, the anemone of Adonis is at the antipodes of ripe fruit. The second feature Ovid mentions affirms the proverbial sterility of the "stone gardens" raised for the Adoniai.[169] Without perfume, without fruit, the anemone is a transient and fragile flower. Weakly fastened and much too light, a breath of air loosens it. It is rootless, vain as the words borne on the wind that Lucian calls "the anemones of speech."[170] Blocked from the animal transformation that would sanction a destiny symmetrical to Atalanta's, the hunter-seducer who becomes an odorless flower and a wind rose sees himself exiled from the world of perfumes and odors. The final flourish Aphrodite tries to win for him is a poor excuse for a flower that a breath of air can prostrate. By a route that doubles on the botanic plane the rival metamorphosis of Atalanta, the voyage from myrhh to lettuce is ratified. Even in the flower born from his death, Adonis is disabused of all links to the world of desire and banished from the land rich in flowers and sweet odors where Eros blooms.[171] In one version, that adopted by the Alexandrian poet Nicander, lettuce and anemone are even associated in a single sequence where one serves as a ridiculous refuge from the rage of the boar while the other is born from the victim's mortal wound.[172]

To follow the hunt of Atalanta, a lioness to Aphrodite's lover, the analysis returns to the world of odors, from panther to anemone, which the hunter sprung from spices conjures up. For the metamorphosis of Adonis into a rose without scent is not just a proof that the myrrh tree sets up a semantic plane essential to the mythology of the fair hunter, victim of the boar. One could go on—if time had not imposed a decidedly arbitrary limit—to enrich the botanical model recently proposed[173] by giving a place above and beyond cereals and ripe fruits to flowers without scent answering to scents without flowers.[174] It would only be a way of marking the relationship between mythical categories and the explicit thought that leads one of Plutarch's guests[175] to distinguish fresh perfumes from "corrupted" ones;[176] if the latter, which are denounced as fradulent and fake, can only engender artificial delights, the former are likened to the produce of the seasons and in no way differ from the ripe fruits that, like them, are natural and pure.

Having lost the privileges he had during the dog-days, the hunter-seducer can find no asylum in that verdant time when Eros rises "at the very hour the fecund earth covers itself with spring flowers."[177] His ruin is consummate. Is he not guilty of having confused forest and garden, just as the virgin huntress is guilty of having transgressed the space assigned to the female nature? The effeminate hunter belongs no longer to one or the other sex.[178] He is sundered from desire like Atalanta the lioness, and their two opposite paths cross where a unique, ineluctable condemnation awaits them, as if with one sweep of the hands the Greek imagination were exorcising specters subverting the dominant model of male-female relations.

3

Gnawing His Parents' Heads

Voleur de feu chargé
de l'humanité des animaux
mêmes.

Rimbaud   15 May 1871

To live alone, one must be
a beast or a god, says
Aristotle. There remains
a third case: one must be
both at once . . . philosopher . . .

Nietzsche

Emotions ran high at the end of the nineteenth century when the British school of anthropology undertook to make known to the world "some survivals from a savage estate" in the thought and society in which Western Civilization had unconcernedly placed its origins and values. The Greeks who had miraculously discovered reason incarnate in man and who had thus been first to recognize the privileged position of humans in the universe—could they have tasted human flesh and dined on man like the Iroquois or the savages of Melanesia? If myths testify to a past state of society, as Tylor and his disciples maintained, then Thyestes' supper, the sacrifice of Lykaon, and the story of Kronos became so many overwhelming proofs that Plato's ancestors bore a striking resemblance to American Indians.[1]

If contemporary Hellenists are victims of insomnia, it is surely no longer because Plato's great-grandfather was a cannibal. And the problem poses itself in different terms.[2] Apart from an unusual ritual like that of the wolfman, in the course of which initiates would partake of human flesh mixed with morsels of an animal victim, anthropophagy in ancient Greece is essentially "food for thought," be it in myths, religious representations, or political ideologies. Consequently, the decipherment of cannibalism can be accomplished in two ways. The first is a kind of thematic interpretation whose domain would include the series of myths and tales in which the motif of anthropophagy occurs in more or less episodic fashion. Dionysos eaten by the Titans, Tereus and Thyestes gorging themselves on their children, the Theban Sphinx devouring the young boys with whom she copulates, Tantalos and Lykaon offering to the gods a meal of human flesh, Kronos swallowing the offspring born to him by Rhea —all are myths in which cannibalism appears central, but which, as soon as they are held up to the light of inquiry, reveal profound differences one from the other. For the signification of what seems to present itself as cannibalistic behavior depends in each case on the context, which alone can decide its real sense. Two examples will suffice to demonstrate the dilemmas of the first method. As for the case of Kronos: a naive reading of Hesiod could lead one to think that Kronos is a cannibalistic father since he gobbles up each newborn child as soon as Rhea holds it between her knees.[3] But once placed within the context of the myths of sovereignty in which this story occurs, the conduct of Kronos takes on a completely different sense.[4] Like Zeus, who is his homologue in the Hesiodic myth, Kronos is a sovereign god whose destiny is to be dethroned by his son, by a child more powerful than his father. To neutralize the danger, Kronos and Zeus resort to the same procedure: swallowing (katapínein). Kronos does not devour piece by piece his progeny born from Rhea, he swallows it alive until forced to disgorge it under the influence of the drug administered him by Zeus's accomplice, Mêtis. She is the same Mêtis whom Zeus, menaced in his turn with the vision of a more potent son despoiling him of his sovereignty, decides to swallow after having first married her. In this way he appropriates all the cunning intelligence (= mêtis) without which his reign would be as transitory as that of Kronos. Neither of the two are true cannibals. They are sovereign gods who wolf down their adversaries in order to defend or found their power.

The second example is furnished by the myths of Tereus and of Polytekhnos,[5] two versions of one story in which a man eats without realizing it the flesh of his child fastidiously prepared for him by his wife. Out of context this monstrous meal fathers every misinterpretation, including that of the Dionysiac banquet and the feast of raw meat.[6] But an analysis of the mythological context makes it possible to insert these myths into an ensemble centered upon honey and thus to specify the sense of the cannibalism practiced by Tereus and by Polytekhnos.[7] Actually, these two myths are parallel versions of a story that begins with a honeymoon characterized by excess and ends with the transformation of honey into rotten matter and excrement. In Tereus's version, the groom who misuses his honeymoon is first condemned to seduce and rape his sister-in-law, then to devour the flesh of his child, finally to be changed into a hoopoe, a bird that feeds on human excrement. In the version of Polytekhnos, who is the woodpecker, master of honey and bees, an equally excessive honeymoon leads the guilty groom by the same route (rape and cannibalism) to perish in the honey he has been rolled in before being abandoned to the bites of insects and the stings of flies. This torment suits perfectly its victim, whose first mistake was to have wallowed too long in the honey and to have eaten it to excess—which is the Greeks' way of describing the honeymoon and the pleasure in which young newlyweds indulge each other. For the myths of Tereus and of Polytekhnos simply tell how an improper use of honey transforms this food into its opposite, into excrement or rot. The transformation is mediated by a cannibalistic phase that other myths in the same group define as the state previous to the discovery of honey, explaining how men ate one another until the Bee-Woman taught them to feed on honey gathered in the forest.[8] So at the conclusion of a structural analysis, the cannibalism of these myths is revealed as the sign of a regression from honey and at the same time as the first stage in the putrefaction of honeyed food before it turns into excrement in the case of the hoopoe or rotten matter in the case of the woodpecker.

In the absence of a systematic reading of the different groups of myths in which the stories involving anthropophagy belong, another road remains open: to define cannibalism within the Greek system of thought, to situate it in the ensemble of representations a society makes of itself and of others through its table manners. In fact anthropophagy, which the Greeks took as a modality of *allélophagía* ("eating one another"), is an essential term in the food code that represents a privileged plane of signification in their social and religious thought for defining the whole set of relations between man, nature, and the superhuman.[9] There is no choice, then, but to deploy this whole system with the goal of enticing cannibalism from the marginal position explicitly imposed on it by a society that categorically refuses to practice it but, by virtue of precisely the things it can say about it, forces protesting groups or individuals to express their protest by adopting the very eating behavior it refuses. In other words, the definition of cannibalism in Greece is articulated not only within the politicoreligious system of thought but also outside it through the different shapes assumed in Greece by the denial of the city-state and its values. This denial was expressed sometimes by more or less isolated individuals such as the Orphics or the Cynics, sometimes by more or less organized sects such as the Pythagoreans or the followers of Dionysos. Whether they considered themselves an antisystem or a protest against the city-state, these four movements—Pythagoreanism, Orphism, the Dionysiac religion, and Cynicism—constitute a four-termed ensemble each of which reflects a mirror image of the politicoreligious system in which cannibalism is marked positively in some cases, negatively in others.

The politicoreligious system bases its dominance on the sacrificial practice that informs the ensemble of political behavior and determines the eating life of the Greeks. The consumption of meat actually coincides with the offering to the gods of a domestic animal whose flesh is reserved for men, leaving to divinity the smoke of the calcined bones and the scent of spices burned for the occasion. The division is thus clearly made on the alimentary plane between men and gods. Men receive the meat because they need to consume perishable flesh, of which they themselves consist, in order to live. Gods have the privilege of smells, perfumes, incorruptible substances that make up the superior foods reserved for the deathless powers.

Here, then, is a first definition of the human condition. But sacrifice also implies a second, not with respect to gods but with respect to animals. This time the boundary is less clear for a series of reasons: first, because men and animals share the common need to eat and thus suffer together from hunger, the sign of death; then because certain animal species are carnivorous for the same reasons as men; and finally, if gods and men are separate to the extent that burning spices are needed to summon the former to the latter's sacrifices, men and animals, on the contrary, live in such familiarity that at times men evince real difficulty in radically distinguishing themselves from an animal, for instance the plow ox.

In the city, the dominant ideology on the relationship of man to the animal world is conveyed by Aristotle and the Stoic school. They agree in believing that animals exist for the benefit of man, that they are there to provide him with food, furnish his clothing, and help him in his toils. That man uses animals for his own ends, says Aristotle, is a just law of nature.[10] The enemies of vegetarianism will echo him: to renounce the use of animals is to run the risk of "leading a bestial life."[11] So man can sovereignly divide the animal word in two: those animals he protects in the expectation of their services to him and those he hunts for the harm he fears from them. Whether domesticated or wild, animals are always considered as beings deprived of reason with which man can establish no legal bond, given that animals are incapable of "concluding any accords between them in terms of which they would inflict no damage and suffer none themselves."[12] The animal world knows neither justice nor injustice. This essential ignorance founds the most radical opposition in Greek thought between animals and men. Separated from humanity, which lives under the regime of *díkē*, or legal bonds, animals are condemned to devour one another. The reign of allelophagy commences at the antipodes of justice: "Such is the law that the son of Kronos laid down for men: that fish, beasts, winged birds devour each other since there exists not any justice among them."[13] Consequently, the true difference between man and animal, as between man and gods, exists on the level of food, but with the distinction that on the animal side the separation is two-fold, in that it comprises two stages, of which the raw is but the first. "Man is not an animal who eats raw flesh."[14] For all of Greek thought, human food is inseparable from the sacrificial fire, while the least savage of domesticated animals, the herbivores, are condemned to eat the uncooked.[15] In other words, bestiality begins with omophagy (eating raw food) and ends in allelophagy.

Between beasts and gods, man's position is well protected. The whole politicoreligious system sustains it through the daily practice of the alimentary blood sacrifice. But in this fixed form the three-tiered model is neither correct nor adequate. It only becomes so when its dynamic character is realized. The human condition is not only defined by what it is not; it is also delimited by what it is no longer. In the Greek city-state, in which cultural history picks up where mythical discourse about origins leaves off, there arises a two-fold tradition marked by the alternation of the Golden Age and the Age of Savagery. At times—so it is in the myth of Hesiod—men came to eat meat after having known admission to the gods' table; at other times—as in the myth of the Bee-Women—men did not attain their diet of today until they had lived for a long time the life of savage beasts, eating raw food and devouring one another.

The model, then, offers two symmetrical openings, one from above, the other from below. They serve as markers in this conceptual domain of two concurrent orientations whose homology is brought out by the presence of the same mediator at each extremity: Prometheus. In one case, by the invention of sacrifice, Prometheus assures passage from the communal repasts of gods and men in the Golden Age to the meat diet.[16] In the other, by bringing fire and inventing various techniques, Prometheus wrests humanity from savagery and bestiality.[17] Between these two representations the city-state felt no obligation to choose, and thus it gave equal place to both. In sacrificial practice, it implicitly assumes the orientation of the Hesiodic myth; while in the different ideologies centered on its own history, it never ceased to honor the passage from allelophagy to a diet marked by the consumption of meat and bread.

Consequently, within the politicoreligious system of thought, cannibalism is clearly denounced as a form of bestiality that the city-state unambiguously rejects and that it exiles either to the confines of its history in an anterior age of humanity or to the limits of its space among the hordes that compose the world of barbarians. The geographical distribution of savages obeys the principle that omophagy is a less marked form of bestiality than allelophagy.[18] Thus, eaters of raw flesh are to be found even in certain distant regions of Greece, such as Northern Aetolia where the Eurytanes live, men "who speak an utterly unintelligible language and feed on raw meat."[19] True cannibals live far, far away: the Scyths are the ones who eat human flesh as others of their kind thrive on mare's milk.[20] Herodotus names them *androphágoi*: "They have the most savage customs, they do not observe justice, they have not a single law. They are nomads . . . they alone of the people of whom we speak partake of human flesh."[21] This is what Aristotle as a clinician calls "bestial tendencies," and he finds them among the same savage hordes of the Pontic region who have "leanings toward murder and cannibalism." Such "leanings" are at times institutionalized, as among certain tribes that, it is said, supply each other with children to feast upon.[22] All are instances of cannibalism at the confines of the civilized world where Greeks can both denounce as Plato does the survival of a primitive state of humanity[23] and equally well recognize, as do Aristotle's contemporaries, the savage tribe to which the bogey Lamia belongs, that Lamia who comes by night to devour fetuses she has ripped from the wombs of pregnant women.[24]

This sociological representation of cannibalism must be complemented by another one that corroborates the city's radical exclusion of all anthropophagic behavior. In Greek thought of the fifth and fourth centuries B.C., the tyrant, *túrannos*, is a kind of man explicitly defined in terms of the three-part model that underlies sacrificial practice and the city-state's dietary system. Since he is the source of his own power without having received it by allotment or having been compelled to put it back "in the middle" (es *méson*), the tyrant raises himself above others and above the law. His omnipotence is an avowal of equality with the gods. But at the same time the tyrant turns out to be excluded from the community and cast out into a place where political thought no longer distinguishes the superhuman from the subhuman, where the distance between gods and beasts disappears.[25] In the Platonic *Republic*, tyrannical behavior marks the emergence into daylight of savage appetites that are normally roused only in sleep, when the effects of drink elicit the bestial part of the soul, *tò thēriôdes*; it commits incest with its mother, rapes man, god,

or animal without distinction, kills its father or eats its own children.[26] Outside the city and the hierarchic system wedded to it,
man, god, and animal become nothing more than the interchangeable
objects of a desire that invests the tyrant and pushes him to commit
incest and parricide, then inveigles him into endocannibalism. By
consuming his own flesh and blood, the tyrant clearly shows that he
is out of bounds, excluded from society just like the scapegoat ex-
pelled from the city during certain springtime festivals in various
regions of Greece.

In fact, eating human flesh means entering an inhuman world
from which one is not sure to return. The "long-lived" Ethiopians
are the companions of the Table of the Sun and the blessed denizens
of the extremities of the earth. When Cambyses in his excess decided
to subdue them, the farther his armies advanced, the more they came
to lack provisions, so that they were progressively reduced to eating
their beasts of burden, grazing on plants and grasses, and finally
devouring one man in six after choosing him by lot. As crazy as he
was, says Herodotus, Cambyses forthwith renounced the expedition,
so great was his fear that his men would devour each other[27] and
become like savage beasts. A similar story is told of some Phoenician
mercenaries and Libyan dissidents whom a famine had reduced to
eating one another. Their conqueror, Hamilcar Barca, had them tram-
pled by his elephants, since he thought that cannibals "could not
mingle with other men without sacrilege."[28] Herein lies a model for
exclusion, one that has stood the tests of efficiency imposed on it by
certain polemical enterprises aimed at those it was decided to denounce as enemies of humanity. Sleeping with his mother or his
sister, slaughtering a newborn in order to eat its flesh and mop up
its blood with a crust of bread—such things are the crimes imputed
to the first Christians. The Greek apologists Origen and Justin held
the Hellenized Jews of the second century A.D. responsible for these
calumnies, and the Jews themselves later paid for them in turn and
more than once.[29]

Moreover, reversal is a fundamental mechanism in a system of
thought that finds its real meaning in the dynamic interrelation of
four forms of antisystem, each of which inverts for its own profit
the terms of the reference model, each borrowing one of the two
openings at the heart of this model. For the politicoreligious system
can be overtaken from above or from below, in the direction of the
gods or, inversely, from the side of the beasts. Pythagoreans and
Orphics explored the first alternative. The second was adopted by
Dionysism and Cynicism in turn. But no matter how protest inside
the city or renunciation of the politicoreligious world proceeds, each

movement is forced to take a certain view on allelophagy and to define itself in relation to cannibalism.

On the religious plane, Orphism and Pythagoreanism are certainly the best-known forms of deviation. To schematize, these are movements of religious bent that reject the city-state's system of values and dispute its model of a primary relationship between gods and men founded on the division Prometheus made when he reserved the meat for men, separated them from the gods, and deprived them forever of the old common repasts. All the action in Pythagoreanism essentially transpires on the plane of food in terms of a more or less importunate demand for a nonmeat diet.[30] In fact, Pythagoreanism assumes two different attitudes at the same time with respect to the blood sacrifice and the city. In one case, there is the intransigent denial of a sect set up as an anticity. In the other, there is the mitigated denial of a more political than religious group that proposes to reform the city from within. The former reject any form of meat as food; the latter compromise and decide that certain sacrificial victims—pigs and goats—are not properly speaking meat: real meat is the flesh of the plow ox, whose slaughter is the object of a formal prohibition.

Of these two attitudes only the first merits the status of "renunciation" that leads it to constitute itself as an antisystem. For strict vegetarians, every blood sacrifice is a murder and ultimately an act of cannibalism for which their horror is expressed by way of the broad bean. In fact, this legume is at the antipodes of spices, the marvellous food of gods and the Golden Age. Thanks to a stalk without nodes and by virtue of affinities with the rotten, the broad bean establishes the same direct communication with the world of the dead as spices establish with the world of the gods to which they belong through their solar quality and desiccated nature. But in the Pythagorean system of thought, the broad bean is still more. It is a being of flesh and blood, the double of the man at whose side it grows and from whose rotten compost they both feed. As a result, say the Pythagoreans, it is the same crime to eat a broad bean or to gnaw one's parents' heads. Proof of it is produced by a set of experiments known to Pythagorean tradition. A broad bean is placed, for its mysterious cooking, in a pot or closed container that is then hidden in manure or buried in the ground. After a more or less lengthy period of gestation, the bean transforms itself either into female genitals with a child's scarcely formed head attached to them or into a human head whose features are already recognizable. In these experiments the pot is a womb entrusted with the task of revealing the broad bean's true nature. But it can already be discovered by taking a half-eaten or slightly squashed broad bean and placing it in the sun for a moment or two. Immediately a smell arises which is said to be either the smell of sperm or that of blood shed in a murder. The Pythagoreans are explicit: eating broad beans is feeding on human flesh, is devouring the most marked type of meat. To the eyes of these abstainers, the carnivores of the city who eat sacrificial victims oscillate between two forms of bestiality: on one side, Lamia devouring the fetuses she rips from the wombs of pregnant women, and on the other, the cannibal son who eats the head of those who are most precious and most sacred to him. By the act that establishes it as an antisystem and a counter-city, the pure Pythagoreans' sect reverses the political model and reduces it from three terms to two. Allelophagy is no longer relegated to the distant savages. It is right in the city-state amid the men who sacrifice on the altars in honor of the gods. That is what the tradition means when it says that Pythagoras invented the vegetarian diet in order to cure his contemporaries of the habit they had of eating one another.[31]

The Orphics adopt just as radical a position. Their renunciation of the world leads them to proceed to the same brutal reversal as the Pythagorean sect. There is a paradox in Orphic thought. Traditionally, the most important teaching Orpheus brought to men is "to abstain from murders" (phónoi), which means, in the vocabulary of the sects, to reject the practice of blood sacrifice and the consumption of meat. Yet the central myth of the Orphics is the story of a sacrifice whose victim is the young Dionysos and whose practicants are the Titans. They eat the child after having subjected him to a culinary process in which his flesh is first boiled, then skewered and roasted. As long as people obstinately see a kind of communal feast in this myth, the paradox will persist. On the other hand, it disappears as soon as the sacrifice of the Titans and the sacrifice of Prometheus are placed side by side. Then one can perceive that of these two primordial sacrifices, one, Prometheus's, represents the act that founds the relations of men with gods; the other, the Titans', is explicitly conceived in the Orphic imagination as a feast of cannibals whose sacrificial character is brought out by the cooking processes it uses. Its consequences are the punishment of the Titans by the thunderbolt and the birth from their ashes of the human race; so each time men offer the gods a sacrifice, they unconsciously repeat the murder and re-enact the cannibalistic feast first prepared by their distant forbears.[32]

The division made by the Orphics exactly coincides with that of the Pythagoreans': the true cannibals are those who accept the meat diet and do not practice the "Orphic life" (bíos orphikós), thereby neglecting both to purify the divine element imprisoned in man by the Titans' voracity and to bridge the gap between man and gods that was opened up by the blood sacrifice. In these two sectarian movements, which occupy a symmetrical, inverted position relative to the politicoreligious system, cannibalism is marked with the same negative sign it bears in the city-state.

Everything changes with the two other protest movements that fill out the series of countersystems enshrined in the history of the Greek city-state from the sixth to the fourth century. In fact, for the Dionysiac religion and Cynicism alike, allelophagy, insofar as it is an extension of the raw diet, becomes the chief tool of the subversion that these two movements encourage at the expense of the politico-religious system. Both work from within, at the heart of the city itself, one on a religious, the other on a sociopolitical plane.

The position of Dionysiac religion with respect to the political model is clear. While Orphics and Pythagoreans transcend the sacrifice from above, Dionysos's faithful do so from below. By omophagy, by the process of devouring the raw flesh of an animal that has been chased through the mountains and then, still living, ripped to pieces in defiance of the rules in force for the political sacrifice, the frontiers between men and beasts have been annihilated. Humanity and bestiality interpenetrate and are confounded. Mainads and Bacchants give suck to the young of wild animals and at the same time tear to pieces panthers and roebucks. It is as though better to become savage themselves, the Dionysiac worshipers must first soothe the fierce creatures and attain their intimacy to the point of identifying with them. Dionysos the savage hunter is not simply the "eater of raw flesh" (ōmádios, ōmēstḗs). The omophagy he requires of his devotees leads them, like true wild beasts, to indulge in the most cruel allelophagy.[33] Various data from ritual confirm the best-known mythical representations. On Chios, Tenedos, and Lesbos, Dionysos hungers for human flesh: the victim ripped apart in his honor is a man, as in the story of the Bacchae or the Minyads. Possessed by the god whom Pentheus fails to recognize, Agaue captures on the hunt her son who has come to enjoy the spectacle of women struck with madness. Lion cub and bullock—and "hairy like a savage beast"[34] besides—Pentheus is, to his mother, a prey she tears apart with her hands and prepares to devour.[35] The daughters of Minyas know her desire. Seated at their looms, they prefer to marry rather than go out into the countryside as Bacchants. Dionysos forces them to join his Mainads, and their madness gives them "the taste for human flesh." They choose one of their children by lot and dismember him like a tender animal.[36]

In all these traditions, cannibalism, having been integrated into the logic of omophagy, seems the consummation of the savage state Dionysism strives to attain. Tasting human flesh and surrendering oneself totally to allelophagy both belong to a set of behaviors designed to make man a wild creature and allow the establishment, through possession, of more direct contact with the superhuman, in this case Dionysos eater of men. One is still entitled to ask whether eating human flesh takes on the same positive value in Dionysiac religion as consuming the raw meat of an animal captured in the hunt. In fact, according to one of the versions of the story of the Minyads, the slaughter of the child ripped apart by his mother provokes an indignant reaction on the part of the other Bacchants. They suspend their chase through the mountains to turn against the unnatural mothers.[37] Likewise, at the end of the *Bacchae*, the death of Pentheus is presented as a murder resulting in defilement and forcing Agaue into exile.[38] This time Dionysos in person pronounces the sentence, thus condemning behavior he himself seems to have dictated. It is true that in both contexts the "eater of raw flesh" turns into cannibals only those who resist him and refuse to acknowledge their wildness.

Even so, devouring human flesh is not alien to the Dionysiac religion, but rather an essential and ambiguous component of it. This is singularly clear from the story of the *Bassaroi*, the followers of Dionysos in Thrace: "Not content with being adepts at Tauric sacrifice, the ancient Bassaroi went so far in the delirium of human sacrifice as to devour their victims." And Porphyry, who relates these customs in his treatise *On Abstinence*,[39] remarks that the Bassaroi "were doing as we do with animals since we also begin by offering first fruits and use the rest to feast upon." But there the comparison ends, since, Porphyry continues, "who has not heard tell how they were struck with madness and hurled themselves at each other, biting themselves and actually devouring the bloody flesh, and how they did not cease to act in this manner until they had exterminated the family of those who first indulged in such sacrifices." Here cannibalism discovers its internal contradictions. It rears up in Dionysiac rite as the extreme form of savagery offered by the god to his faithful. But this time Dionysos does not intervene to condemn the eaters of human flesh. They destroy themselves. Seized by a voracious hunger, the Bassaroi are incapable of sating themselves with just the sacrificial victims. They hurl themselves at one another, tearing themselves apart in fury and devouring each other like savage beasts. Allelophagy is at its peak. But—and therein lies its ambiguity—it cannot reach its culmination without condemning itself. One of two things happens. Either the Bassaroi kill themselves to the last man, or the cannibalism endemic in Dionysiac rite is subjected to severe restrictions. To put an end to the epidemic of anthropophagy that threatens to exterminate the group of Dionysiac faithful, the Bassaroi are forced to kill those among them who are known to have introduced these horrendous sacrifices. In its pure form, cannibalism is untenable. The very religion in which it is inherent and which integrates it into certain of its rituals can only make controlled use of it. This is precisely because the Dionysiac movement remains an essential component of the political religion it aims to transcend by way of savagery. To be sure, Dionysiac religion contests the official religion, but it always does so from within, never taking on the form of an antisystem radically alien to the city-state.

Cynicism takes on this impossible kind of cannibalism for a scheme apparently parallel to the Dionysiac one since it likewise aspires to a return to primordial savagery.[40] But though the Dionysiac movement is chiefly religious, Cynicism is a form of thought that attacks the whole of society. In their theory and daily practice, the Cynics actually put into question not just the city, but society and civilization as well. Their protest is a generalized critique of the civilized state.[41] It arose in the fourth century with the crisis of the city, and one of its major themes is the return to a savage state.[42] In negative terms, it is the denigration of city life and the refusal of the material goods produced by civilization. In positive terms, it is an effort to rediscover the simple life of the first men who drank spring water and fed on acorns they gathered or plants they reaped.[43] To learn to eat raw plants once again, the Cynics provided themselves with two models: the savage peoples who had kept this mode of life unchanged and animals, who had never been corrupted by Promethean fire. In fact, Cynicism is riddled with an anti-Promethean strain opposed to the invention of civilizing, technological fire.[44] To grow wild it is not enough to eat raw food or to practice omophagy as Diogenes did when he fought some dogs for a piece of raw octopus (it cost him his life).[45] It is also necessary to deconstruct the system of values that founds society. The return to a savage state proceeds by the critique of Prometheus, who is no longer the sacrificer responsible for the separation of gods and men, but the civilizing Titan of cultural anthropology, the mediator guilty of having led humanity from the savage state by giving it the poisonous gift of fire.

Dionysiac religion throws itself headlong into savagery, seeking possession and contact with the superhuman. In Cynicism the procedure is altogether different. One enters savagery backward; descent into it is progressive and gradual. What begins as eating raw food and condemning fire ends in the enunciation of two basic demands:

abolition of the incest tabu and the practice of endocannibalism. Both of these were formulated by Diogenes of Sinope, the first in an ironic mode. Why does Oedipus so violently lament being at once father and brother of his children, at once husband and son of the same woman? Roosters, dogs, and donkeys don't make so much noise about it, and neither do the Persians.[46] As for the second demand, Diogenes did not content himself with his discreet expression of it in the observation that more than one society had not forbidden itself to taste human flesh. It is said that he himself "taught children to push their fathers and mothers to the sacrificial altar and to eat them to the very last bite."[47] Incest, parricide, and cannibalism: the big tabus are cast down. Society is deconstructed down to virgin soil, where Cynicism finds nothing more than the individual in his pure state, before society, and on the hither side of all forms of group life. Only in the context of such radical questioning of the civilized state can cannibalism take the full positive value that the Dionysiac movement, as a collective one integrated into Greek society, could not acknowledge with such candor or, for that matter, so gratuitously. For it is all too clear that only "intellectuals" like the Cynics can allow themselves to eulogize the cannibal son in order to affirm the right of the individual in defiance of society and every form of civilization.

These four solutions with the politicoreligious model at their center can be distributed in pairs each of which affixes either a positive or negative sign to cannibalism. The first two pose as antisystems and are content to reverse the political model, condemning within the city-state anthropophagic conduct that it itself condemned without. As for the other two, they proceed from within the politicoreligious system, undertaking to unleash cannibalism upon the city either to destroy it or to introduce within it what Plato would call the Other. Even so, these four solutions do not exhaust the possible combinations opened by the model with which we began. In fact, this model actually produces only two solutions, each of which appears in two different modalities in the course of history from the sixth to the fourth century. Nevertheless, there are several reasons for speaking of them as a system. First, there is the constant alternation of the Golden Age and savagery in the mythical and legendary tradition of origins. Then there is the fact that Prometheus changes complexion depending on whether he has assured mediation since the common repasts of men and gods or since the savage state men shared with animals.

To these two arguments history adds another, which confirms the system's overall coherence and the relationship between its two symmetrical openings. It consists of the transformation of one solution into its opposite, and we can observe it taking place in the second third of the fourth century, when certain Pythagoreans, after their failure in the political and religious spheres, transform themselves into Cynics almost before our eyes.

At the end of the fourth century, Pythagoreanism had definitively miscarried in its political goal of reforming the city. Hunted and persecuted, the disciples of Pythagoras dispersed. A small group succeeded in surviving at Phlious; Arkhytas departed to pursue his career at Tarentum; the majority abandoned Magna Graecia.[48] Pythagorean society was dismembered, neither sect nor brotherhood remaining. A certain number of refugees met at Athens. Their character, appearance, and style of life are known to us from comedies written for the Athenian public by poets like Antiphanes, Aristophon, and Alexis.[49] By surprise, the Pythagorean has become a comic figure. Nothing remains to recall the solemn person robed in white who gave himself utterly to ascesis and practiced sainthood within the harsh limits of the sect. The new Pythagorean is a hobo. Barefoot and covered with dirt, he quenches his thirst in the gutters[50] or climbs down ditches to graze on weeds like orach.[51] A beggar's wallet slung over his shoulder, a coarse, short cloak on his back, he sleeps under the stars summer and winter. In short, there's no end to the laughing. But these hairy, filthy vagabonds—can they still be called Pythagoreans when nothing in their behavior seems to distinguish them from Cynics? They have even borrowed the Cynic's special tokens, his wallet and short cloak.[52]

That question has been posed by historians of philosophy since antiquity, and their perplexity is largely shared by moderns.[53] In Athens during the first half of the fourth century there was a strange philosopher named Diodoros of Aspendos who claimed to be a disciple of Pythagoras but dressed and behaved like a Cynic. In the words of one of his contemporaries, he is "a follower of Pythagoras, but one who gathers a large crowd round his beast-robed madness and insolence."[54] Is the confusion so serious that we should follow the obscure author of a book called *The Philosophers One after Another* and accuse Diodoros of vanity in affecting long hair and a short cloak "though the Pythagoreans before him dressed impeccably, rubbed on ointment, had their hair cut and beards trimmed like everybody else"?[55] Instead, Diodoros's ambiguous character personifies the mutation attested to by comic poets of the same era who exploited such figures for their surest laughs: nothing funnier than a stageful of strict vegetarians fighting hungrily over the remains of a dog.[56]

Everything seems to happen as if, once the solution from above had met with failure, the last Pythagoreans had only one way out or the other: either return to the city and disappear or try, as individuals, to defeat the politicoreligious system from below. Perhaps the "gods-men-beasts" model gives the most satisfactory explanation of the way the same dissenters could, within the lapse of two centuries, first affix a negative, then a positive, sign to the image of a child devouring his parents.

4

The Orphic Dionysos
and Roasted Boiled Meat

If Dionysos has long borne the guise of foreigner—though, as we know today, his stock is as good as that of other Greek divinities—the reason lies in his affinity for what is foreign. At once present within and without the city, Dionysos delights in the ritual game of hospitality, *xenismós*, in which this citizen of the Pantheon and certified Olympian is received and welcomed by the whole political community as an alien power. In the enclosed space of the city and beyond it, he calls up at will the haunting figure of the Other, wearing a mask that exposes but always conceals him, particularly when he seems to show his most familiar face. Thus, on a given vase Dionysos assists at the initiation of one of his faithful whose appearance precisely matches his own, just as the Dionysos called *Mústēs* initiates himself in his mysteries and thus divides his own otherness in two where mysticism opens a space for transgression.[1]

Dionysos does not cease being evasive and protean in the interpretations he provokes, today as before. At the beginning of the century, in a cold war climate, a whole school of historians undertook to discover "the mystery religions" and their relation to primitive Christianity.[2] Dionysos, whose power was at the center of the debate, remained ever himself, keeping the same smile for fanatic devotee and blind disdainer. Was he not the god who saves by initiation in his mysteries, who delivers his faithful, be they women or slaves? Did he not appear in the "communal feast" of his initiates as the model of those gods who offer themselves as victims in the paroxysmal type of sacrifice? For V. Macchioro, one of his most ferocious defenders, who tried to show in a series of studies how Pauline Christianity transposed the Orphic tale of the passion of Dionysos, there wasn't the least doubt that in shedding his humanity the Orphic became Dionysos just as the Christian became Christ.[3] On the contrary, A. J. Festugière's preoccupation is with the demarcation of Hellenistic religious fact at its most dangerous frontier, the problematic intersection of Christian mystery and what the history of religions then called "the pagan mysteries." For him, the so-called passion of Dionysos is reduced to a late legend plagiarized from the pitiful death of Osiris, and the mysteries of this god could only offer his poor initiates a few hours' respite from the tedium of their daily existence.[4]

Interpreted by Neoplatonists and the Church Fathers as well, the Orphic myth of Dionysos slain and devoured by the Titans has not ceased to raise the basic questions of blood sacrifice. For historians like R. Smith, J. Frazer, and A. Loisy,[5] the passion of a god placed at the confluence of the ancient Near Eastern religions with the new Christian doctrine of salvation seemed to formulate such questions with the utmost pertinence. Inseparable from ancient agrarian rites, Dionysos seemed equally inseparable from the new orientations of religious thought. Doesn't one find, right in the *Bacchae* of Euripides, the "already almost Christian"[6] idea of a purely spiritual ecstasy that makes life with Bacchus a holy ceremony? The fortune in Orphic thought of a god whose "passion" became an exemplary text for the ideology of sacrifice only accentuates Dionysos's marginal character. It is as if being accepted in a sectarian milieu condemned the Dionysos slain by the Titans to be interpreted only on the periphery of his own mythic space.

To escape the whirlwind of these ideologies in hermeneutic guise, a firm anchor in the givens of the tradition is necessary. First of all, this myth relates not only the slaughter of young Dionysos, but also the punishment of those responsible, the Titans, whom Zeus's thunderbolt reduces to ashes from which the human race will rise. Furthermore, this story, which is contemporary with Peisistratos, holds a central position in the oldest Orphic teaching. It dates from the sixth century.[7] From then on the myth of Dionysos slain by the Titans becomes,[8] in its various sequences, the source of a series of questions about the alimentary blood sacrifice, its modalities, and its status in Greek thought and society. These questions spring from the double enigma posed by the unusual aspect and the monstrous character of the sacrificial murder committed by the Titans. Monstrous, for covered with gypsum and masked in plaster, the Titans captivate a child, the young Dionysos, by disclosing to him beguiling objects—a top, a rhombus, some articulated dolls, knucklebones, a mirror. And while Dionysos contemplates his own image caught in the shiny metal, the Titans strike him, cut him up, and indulge in such a strange kind of cookery by the standards of Greek culinary tradition that an entire Aristotelian *Problem* is devoted to it.[9] The victim's limbs are tossed into a cauldron where they are brought to a boil, after which the Titans grab them, skewer them on spits, and set them to roast.[10] And they have the time to devour this roasted boiled meat, all of it except the heart, which is hidden and thus preserved from the destruction which Zeus's thunderbolt inflicts on the executioners.

But this human sacrifice is not only puzzling for the procedures it describes, it is still more singular in the central position reserved for it by the Orphic movement, one of whose central tenets is the obstinate refusal in the dietary practice of its way of life to eat meat, or, in religious and cultic terms, to offer blood sacrifices to the gods.[11]

This sacrificial paradox, once it is explained, must be our point of departure. What does this contrast between the myth and the practice of the milieu formulating it signify? Like Pythagoreanism, its contemporary, Orphism is a movement of religious protest that defines itself by an attitude of refusal, refusal of the whole politicoreligious system organized around the Olympian gods and the distance that separates them from men.[12] In Greek thought, the source of this differentiating split is the sacrifice of Prometheus, which specified the parts of the animal victim reserved for each: for the gods, the savor and smells they are content to sniff, for the men, the meat they need to live and which is the sign of their mortal condition. To change one's diet is to throw into doubt the relationship between gods, men, and beasts upon which the whole politicoreligious system of the city rests. But the refusal takes on different modulations depending on which of the two, Orphics or Pythagoreans, is refusing.

For the Pythagoreans, rejecting the city can take on two forms, which translate into parallel dietary behaviors. Either the refusal is pure renunciation of the world, in which case the style of life chosen is an ascesis, a complete purification of the soul—any practice of blood sacrifice is prohibited, and meat as food is rejected with intransigence; or the critique of the politicoreligious system proceeds from within and is reformist. Then Pythagoreanism is a political movement of religious inspiration that aspires to transform the city-state. This orientation results in different dietary behavior, as evidenced by the Pythagoreans, who accept pork and goat meat as food but who obstinately refuse to touch beef or mutton, as if these two animal species alone represented meat. Pigs and goats are excluded for various reasons specified by Pythagorean ideology.[13]

For the disciples of Orpheus, on the other hand, only one attitude is possible since Orphism is an exclusively religious movement. It is a sect that radically questions the official religion of the city-state and at two levels in particular. The first is that of theological thought, the second that of practice and behavior. Basically, Orphism is a book religion, or rather, a religion of texts, with cosmogonies, theogonies, and the interpretations they ceaselessly generate. All this literature seems in essence built up to counter the dominant theology of the Greeks, that of Hesiod and his *Theogony*. Three examples will suffice to show the contrast between Orphism and Hesiod.[14]

The first is the opposition between Hesiod's Chaos and the primordial Egg of the Orphics. The Hesiodic *Theogony* places at the origin of all things a power of the "un-organized," the Gap, Chaos. With it as a point of departure, the constitutive powers of the cosmos will distinguish themselves, take shape, and define themselves in relation to one another. The sovereignty of Zeus marks the end of a process that starts with nonbeing and concludes with being. In the Orphic cosmogonies, the direction is reversed. At the origin of everything is the Egg, symbol of life, image of perfect being. And the Egg, which represents original plenitude, will slowly crumble into the nonbeing of individual existence. A second example: Eros, Love, who plays a principal role in the *Rhapsodic Theogony* of the Orphics, is, so to speak, put in parentheses by Hesiod. Among the former, Eros assumes the names of First-Born, *Prōtógonos*, and of that which makes things gleam, *Phánēs*. He is a power who integrates, who reconciles opposites and contraries, the primordial force that makes it possible to unify the differentiation in a world torn by the action of *Neîkos*, Strife. In the *Theogony* of Hesiod, on the other hand, Eros is not more than the principle of generation by copulation whose mediation makes possible the differentiation of distinct powers. The divergent orientation of these two systems of thought appears still more clearly in the place each reserves for man and the problem of his generation. For Hesiod, true discourse on being is the gods, their respective lots, their history through the definitive victory of Zeus. The division marked off by Prometheus's sacrifice only ratifies, on the level of myth, the order defined by divine powers. Conversely, in Orphic thought, the generation of man is a capital chapter: it explains how the first men appeared in an originally perfect world and why they were condemned to individual existence while retaining a particle of divine origin.

Since Orphism consists of a body of writings inseparable from a way of life, its rupture with official thinking results in equally great differences in terms of practice and behavior. The man who chooses to live in the Orphic way, the *bíos orphikós*,[15] presents himself, to begin with, as a marginal individual. He is a wanderer like the *Orpheotelestaí*, who go from city to city proposing their recipes for salvation and walking over the earth like the craftsmen of yesteryear.[16] To be sure, these quasi-monks are not only cut off from the political world of the city, they deliberately shun it. They even took care to mark the distance between themselves and the city in practical terms. Their clothing stood out, for they wore white garments only, and they refused to let themselves be buried in woolen garments since wool is also completely alive.[17] But above all, their way of eating identified the disciples of Orpheus. For Aristophanes' contemporaries, one phrase sums up the whole doctrine of the sect: "Orpheus taught men to abstain from *murders*."[18] The "murder" in question is in Greek *phónos*, a word familiar in sacrificial vocabulary.[19] Whatever some may say,[20] Orpheus did not come to teach men respect for human life. The city-state was already busy with that. Orpheus's originality lies elsewhere, in his extension of the concept of "murder." There is *phónos* as soon as that which is animate is killed, as soon as the living being is destroyed. And for all the Orphic thought of the sixth century, to abstain from murder is essentially to refuse the blood sacrifice and the meat diet inseparable from it. The so-called Orphic way of life is not reducible to an insipid vegetarianism. To abstain from eating meat in the Greek city-state is a highly subversive act.

Such is the cultural and religious backdrop of the story of the death of Dionysos told by the disciples of Orpheus. This is a myth about the blood sacrifice, and it stands at the center of a system of thought that rejects this kind of sacrifice and establishes itself in open opposition to official tradition. The tradition's representative is Hesiod, the theologian who tells of the decisive division of food at the first animal sacrifice when Prometheus assigned to each of the partners—to men and gods—his definitive place. As a consequence, the Orphic myth of Dionysos sacrificially eaten by the Titans can only be deciphered with reference to the Promethean myth of the first animal consumed. Such a confrontation is the more justified since in both cases, as we will see, the central concern is a blood sacrifice taking place at the origin of things and defining at once a way to eat and a type of relationship between gods and men.

Among interpretations of the slaughter of Dionysos, one of the most stubborn would see in the deeds of his murderers an enactment of omophagy, the specifically Dionysiac rite in which the flesh of an animal victim is eaten raw.[21] The ideological motivation therein is plain: the myth ought to reflect a ritual. The reflection is hardly muddled by the inversion of Dionysos who, once a sacrificer unleashing the violence of his Bacchants in the *diasparagmós*, becomes in this tale the sacrificed victim of a raging band of Titans. For a whole series of historians with Frazer at their head,[22] the preeminence of this ritual was all the more convincing since it was an integral part of an agrarian religion whose permanent presence in the most diverse civilizations the Mannhardt School was enraptured to discover. To this so confident interpretation, it suffices to oppose one single objection, a decisive one: the Orphic myth of Dionysos makes no allusion to omophagy or to the raw food of Dionysiac religion. No version of this myth exhibits the Titans capturing the child Dionysos at the conclusion of hunt or chase. Moreover, none tells us that the Titans devoured the flesh of their tender victim raw.[23] In fact—and at this point we enter into the analysis of the myth by its most pertinent traits—the story of Dionysos and the Titans falls unambiguously into the category of cooked and cooking. The butchered victim is carefully boiled before being roasted; we are far from a lynching scene.[24] The narrative has two features of unequal significance that conspire to qualify the Titans as performers of a regular sacrifice perfectly homologous to that of Prometheus.

The first is the way the Titans treat their victim. They do not track him, hunt him, or do him any other violence than the inevitable cutting of his throat. In the ordinary sacrifice, the rule is calmly to lead the victim, a domestic animal, to the place chosen for sacrifice, and then to try to obtain its consent either by sprinkling it during the libation or by pouring grains of barley and wheat in its ear.[25] The animal is thus persuaded to shake its head in a certain way, which the Greeks interpreted as a form of assent.[26] The Titans aspire to the same purpose by offering Dionysos marvellous toys: articulated dolls, golden apples, rhombus, top, and, finally, the mirror[27] in which the child finds his own image and loses himself in its contemplation.[28] All are tricks to enable the Titans to strike their victim without his noticing or struggling or letting out an inauspicious shriek. But is the weapon of the crime really the sacrificial instrument? Only the Dionysiac epic of Nonnos breaks the silence of the other versions. It tells us that Dionysos, absorbed in the contemplation of his face in the distorting mirror, was slain, killed with a knife, the *mákhaira*.[29] In the *diasparagmós*, the victim is torn to pieces by bare hands or, as in the sacrifice to the *Déspoina* of Methydrion in Arcadia where the priest did not slit the animals' throats, the crowd hurls itself at the victims and each person ripped off whatever piece he could grab.[30] Here, on the contrary, the Titans use the sacrificial knife, the instrument capable of slitting the victim's throat and also cutting the victim into pieces. All the other *testimonia* tell us only that the child was cut up or dismembered, and the verb used to express the dismemberment, *diaspáō* or *diasparássō*,[31] has often led to the conclusion that this is a scene of omophagy. But the rest of the story, which is in every case concerned with the details of the cooking procedure, forbids us to recognize the practice of *diasparagmós* in this episode. If the narrative resorts to a Dionysiac term of such violence, it does so in order to leave no doubt as to the Titans' brutality. Dionysos is slaughtered savagely by sacrificers who are also executioners.[32]

The second feature is without doubt more significant. Traditionally, boiling and roasting belong to the prerogatives of the sacrificer-cook.[33] With the cooking at Dionysos's expense, we are squarely ensconced in the economy of the normal sacrifice. Moreover, most versions of the myth begin with the narration of these operations. Cooking the victim's flesh is an essential part of the sacrificial ritual. But in the case of the Titans,[34] this culinary process is more complex than usual, to such an extent that it is the subject of an Aristotelian *Problem* centered on the relationship of boiling and roasting: "Why is it not permitted to roast boiled beef, while it is permitted to boil roasted beef? Is it because of what is said in the *Teletē*[35] [the Orphic narrative *Initiation Rite* in which the myth of Dionysos and the Titans was told]? or is it because men only learned later on to prepare boiled food, since in the old days they roasted everything?" There is the problem: there are two ways to cook meat, roasting and boiling. Their order of execution is not indifferent. However, it seems true, according to this *Problem*, that the Titans reversed the normal order. They began by boiling Dionysos's flesh before putting it on the spit.[36] Why? In fact, two questions arise: what is the relationship between roasting and boiling in the ordinary sacrifice? What is the sense of this relationship and of the inversion it can be made to undergo?

On the first point, the Aristotelian *Problem* already clearly indicates that the two ways of cooking cannot be confused and are not interchangeable, as is commonly said and as the custom (attested more than once) of simply roasting all the meat could lead one to believe.[37] In fact, a whole set of information—literary,[38] ceramic, and epigraphic—formally demonstrates, first, that the economy of sacrifice comprises two distinct cooking processes: cooking in the cauldron and roasting on the spit; and second, that these two ways of cooking apply to different parts of the animal sacrificed and always succeed each other in the same order. The viscera are first roasted on spits, then the rest of the meat is set to boil in the cauldron. In the great inscription that defines the rules of the brotherhood of the *Molpoi*, the Bards of Miletus, a family of priests is commissioned to "broil the viscera and cook the meat by boiling."[39] This division of the culinary task into two ritual periods is remarkably illustrated by an Ionian hydria from Caere on which the various phases of a blood sacrifice are represented.[40]

This opposition between boiling and roasting is ratified within the sacrificial model by the game of opposites among parts of the victim. Some are on the side of the spit, the rest belong to the cauldron.

The parts called *splánkhna*, viscera, are reserved for the spit. Aristotle draws up an exhaustive list of them in his *Treatise on the Parts of Animals*: liver, lungs, spleen, kidneys, and heart (stomach, esophagus, and intestines are excluded).[41] These constitute the internal organs, which the Greeks defined by opposition to what they called meat, *sárx*, the external parts.[42] This within/without opposition combines with another, more important one between vital/nonvital. The viscera, *splánkhna*, represent an animal's vital parts, as a function of which Aristotle distinguishes in the same treatise two categories of animal species: sanguine and nonsanguine. Only the former possess viscera because the *splánkhna* are organs formed from blood: "a bloody liquid thickens and congeals to form the viscera."[43] This is an essential datum for understanding why these parts of the animal are eaten first. The *splánkhna* represent what is most alive and precious in the animal victim offered in sacrifice. In certain legendary or mythical narratives, power or victory belongs to the one who gains possession of a victim's vital parts. In a Greek myth told by Ovid, an oracle made it known that whoever was the first to eat the *splánkhna* of a monstrous bull confined in the coils of the River Styx would become strong enough to triumph over the gods and overthrow the power of Zeus. At the moment when the giant Briareus was about to roast the victim's entrails, a kite snatched them from him and placed them in the hands of Zeus, thus dispelling in the nick of time the danger that menaced the Olympian order.[44] The same kind of thievery appears in the legendary history of Veii, the Etruscan city besieged by Camilla. Beneath the town the Romans dug a tunnel that emerged just a step from the temple of Juno. They appear right in the middle of a sacrifice at the moment when the augur, after having examined the entrails, announces that "the god grants victory to the one who proceeds with the sacrifice." At once Camilla's soldiers obtain the *splánkhna* and hand them over to their leader."[45]

Other oppositions complement the previous ones. For example, although the victim's flesh is to be eaten salted, the *splánkhna* must be consumed without salt. This prescription may appear gratuitous, but it takes on its full meaning when placed in the context of culinary concepts provided by a comedy of Athenion written in the third century B.C.[46] In this play, a cook who is also a sacrificer—the word *mágeiros* has both meanings—eulogizes his art. Before the invention of the culinary art, humanity was condemned to savagery. Civilization begins with cooking, and all progress of one is a victory for the other. For this philosophical cook, the history of man begins with preculinary allelophagy and culminates in the discovery of refined eating, which marks the flowering of civilized life. And in this history the consumption of unsalted *splánkhna* is adduced as proof of an intermediary state between cannibalism and the discovery of season-ings, *hēdúsmata*, whose use heralds the arrival of stews and abun-dantly spiced dishes. The salt-free skewers of viscera eaten in the sacrificial ritual thus play the same role in the theory of Athenion's spokesman as the "whole grains," the *oulókhutai*, in the history of civilization Theophrastus undertakes to reconstruct in the same era. He likewise uses the sacrificial ritual as his data.[47] "Whole grains" are thrown on the ground before oneself and the victim in memory of a time when the mill had not yet been disclosed to men by Demeter. The *oulókhutai* preserve the vestiges of an intermediate age between the appearance of cereals and the invention of milled wheat and bread made from flour.[48] Consequently the opposition salt/salt-free seems to depend on a kind of periodization of cultural history inscribed in the sacrifice and the detail of its cooking ceremony.

But it is not the sole sign of temporality presented by this ritual. A new opposition between the *splánkhna* and the rest of the meat shows that the sacrifice incorporates a foretime and an aftertime. A set of cult rules like the calendar of Erkhia distinguishes two phases in the sacrifice as it unfolds: until the *splánkhna* and after the *splánkhna*. The sacrifice in question is made to Zeus Meilikhios at Athens in the month of *Anthestēriōn*. The victim is a ewe, and it is specified that the sacrifice will be of the type *nēphálios mékhri splánkhnōn*, i.e., there will be no wine used until the *splánkhna*.[49] In other words, after the skewers of *splánkhna* are eaten, wine will be permitted as it usually is for libations at an ordinary sacrifice. This ritual detail is especially pertinent since it underlines the double aspect of the cult of Zeus *Meilíkhios*, a half-chthonic, half-celestial god who is by turns malevolent and beneficent.[50]

These various oppositions are not unrelated to one another. If eating the viscera necessarily constitutes the first phase of the sacri-fice, this priority is doubtless not foreign to the vital nature of the organs called *splánkhna*. The viscera are eaten first because they assure maximum participation in the sacrifice. Two witnesses suffice to establish this. The first is a scene in the third book of the *Odyssey*. At the moment Telemakhos accompanied by Athena comes in sight of Pylos, Nestor and his sons are busy offering solemn sacrifice to Poseidon. "Already they had partaken of the first grill, the *splánkhna*, and they were burning the thigh-pieces, the *mēría*." Once the people of Pylos catch sight of Telemakhos, they rush up to invite him. But instead of contenting themselves with offering him a piece of the portions of meat no one had yet touched, the Pylians suspend the proceedings. They beg their guests to pour a libation, to utter a short prayer, and to eat a part of the *splánkhna*.[51] In this way, Telemakhos and his comrades are fully associated with the sacrifice. By receiving their share of the viscera, they take their place in the circle of those the Greeks called "eaters of *splánkhna*," *splankhneúontes* or *sus-splankhneúontes*, "those who eat the *splánkhna* together."[52] The solidarity of these co-eaters of viscera is explicitly attested by the second witness, a ritual of the Eupatrids designed to purify a person guilty of a crime of blood. This ritual opens with a sacrifice in which an appointed purifier takes part along with a series of personages whose names are not disclosed but whose complicity with the purifier is confirmed by sacrificial actions: they all eat the victim's vital parts together, and this sharing of food qualifies them to accomplish the murderer's purification collectively.[53]

All this leads one to single out one opposition as dominating the rest, the opposition between two circles of eaters of sacrificial meat. The first is the relatively tight circle of co-eaters of *splánkhna*; the other is the larger, less restricted one of participants in the sacrificial meal. Between the two circles lies not only the distance separating roasting and boiling but also the difference between two ways of eating. The victim's viscera must be eaten on the sacrificial premises, while the portions of meat can be eaten later, either on the premises nearby or in the private dwellings of those who by participating in the sacrifice have benefited from the distribution of the meat, or, finally, in more or less distant places, among those who have received from the sacrificer the honor (*géras*) of a choice piece of the victim. In this way, for example, at a solemn sacrifice offered to Apollo in the city of Khaleion in Western Lokris a piece of meat was put aside to be sent to a poetess of Smyrna whose poem in honor of their god had been much appreciated by the citizens of Khaleion.[54]

There are, then, a series of contrasts between *splánkhna* and non-*splánkhna*. The former are the internal parts of the victim, its vital organs, which are eaten first on the premises of the sacrifice itself. Eaten without salt, they forge a strong bond between those who share them. As for the non-*splánkhna*, the rest of the meat, it consists of the external, nonvital parts, which can be prepared with salt and seasonings. But their consumption can be deferred and does not result in the same degree of bonding. These various oppositions overdetermine the initial split between the cauldron and the spit, whose complementarity governs the order of each alimentary blood sacrifice, and they confirm the order of the relationship between roasting and boiling in a ritual in which the viscera are always grilled on spits and eaten before the rest of the animal. Boiling always comes after roasting.

Having identified this culinary model of the sacrificial ritual, it remains to discover, first, the signification of this two-fold culinary process—what boiling means in relation to roasting—and then the meaning of the inversion of this model perpetrated by the Titans in the course of their sacrifice. The second problem will lead us to a closer look at the actors involved in the Orphic drama.

The Greeks explicitly remark on a series of differences between roasting and boiling, some culinary, some cultural, without there being discontinuity between the two levels. First, there are qualitative judgments with reference to a physics of cooking. Roasted meat is rawer and thus drier than boiled meat. Again, grilled meats are drier outside than inside, while the opposite is true for boiled meats. Aristotle's *Meteorologica* or the *Problems* of his school are filled with hackneyed observations of this kind.[55] But other differences are less immediate, and in them culinary intermingles with cultural. For example, when Philokhoros compares the respective merits of roasting and boiling, he does not just explain that the spit is less beneficial than the cauldron since boiling not only rids meat of its rawness but also softens its hard parts and cooks the rest to a turn (*pepaínein*). Philokhoros expands on this. He avers that roasting most often varies between two kinds of bad cooking, half-raw and half-burnt, while boiling represents the most perfect kind of cooking, for it recapitulates in culinary art the process followed in the ripening of fruit, the complete *pépansis* that nature obtains by combining dry and moist through the regular interplay of seasons alternating with the rhythm of the sun.[56] With boiling, the art, or *tékhnē*, of cooking rises to the level of a natural process that puts at man's disposal the most perfect foods. The superiority of boiling to roasting is not just gastronomic, it is primarily cultural. Plato confirms it in the *Republic*, in which roasting and boiling are mutually opposed as primitive state and refined, civilized society.[57] In the former, the simple life prevails in which men roast acorns and myrtle berries over charcoal. In the other, there are only stews, sauces, and simmered dishes side by side with spices and sweets. The importance of the contrast between these two regimes is enhanced by its reprise in the description of the Guardians' way of life.[58] To resemble watchdogs who are always alert, their ears erect and their eyes peeled, the Guardians of the Republic will be confined to the roasted meat diet, *optà kréa*, since roasting demands little preparation and none of the tools necessary for stews and meats cooked in pots. To be sure, Plato is here taking up the cudgels against common belief. Praising roasted meat is a way of denying history. But even negatively he attests to the fact that the time of roasting preceded the age of boiling. Such is the

meaning that Athenion or his cook claims to discover in the sacrifice
from the custom of eating skewered meat without salt: the spit is
indeed older than the pot. The same is said by the author of the
Aristotelian *Problem* about the Titans' cooking. One of the interpretations he considers to explain the tabu against roasting boiled meat
is that such a procedure would threaten to subvert a cultural history
persuaded that humanity ate grilled meat until it learned the art of
simmering stews.[59]

From the convergence of these various witnesses one can doubtless
conclude that in Greek thinking the sacrifice, inasmuch as it adheres
to the rule of spit first, then cauldron, incorporates a certain representation of cultural history. The features of this cultural history are
more clearly defined for being the extension of others found in various
mythical narratives on the origin of civilized life. Actually, there is
the same distance between roasting and boiling—both modalities of
cooked—as there is between raw and cooked. Just as cooking distinguishes man from the animal who eats his food raw, boiling
separates the truly civilized man from the bumpkin condemned to
roasting. Herein, one should already note, is a schema correcting the
one in the Prometheus myth. In it the blood sacrifice represents
humanity's passage from a Golden Age to a state of ambiguous evil:
eating the cooked flesh of a sacrificed animal means regressing from
a better toward a bad. But right inside this model the analysis re-
covers another with a different orientation: from bad to better, in the
direction leading from roasting to boiling.

Surely this is the place to adduce by way of paradoxical confirmation the evidence of certain of Pythagoras's disciples. One of their
precepts was not to roast boiled meat.[60] Such a suggestion was certainly not formulated by the pure Pythagoreans, who were determined
never to commit the least "murder," but it suits perfectly those whom
a compromise with politicoreligious power naturally led to sanction
the prescribed order of the basic blood sacrifice. Once the sacrifice
is accepted with the modifications known to us elsewhere,[61] there is
no question of destroying it from within. Far from being a distorted
echo of the Orphic myth, the Pythagorean precept seems instead
designed to differentiate once more the disciples of Pythagoras from
the faithful of Orpheus.

Consequently, the inversion the Orphic myth inflicts on the sacrificial culinary procedures is no longer self-explanatory. Its signification only emerges with reference to the model we have just extracted.
To go from boiling to roasting is to effect a decisive reversal that
effaces any positive aspect in the operation of the ritual itself.[62] But
to interpret this work of denial correctly, it is first proper to cross-

examine the actors chosen by the Orphics to execute this primordial sacrifice. Who are these Titans? Are they simply princes, powers with a calling for sovereignty who are distinguished by an arrogant violence that Dionysos, after Zeus, is forced to suffer for? In the image disclosed to us by different versions of the myth, two details stand out: first, the Titans who get hold of Dionysos are creatures covered with gypsum.[63] In addition, their ashes, mingled with the earth, will give birth to the first representatives of the human race.[64]

The disguise assumed by Dionysos's killers aroused the interest of H. Jeanmaire, eager as he was for any mythical or ritual data from ancient Greece that ethnology could clarify with comparisons. Does not this band of masked men covered with plaster surprisingly resemble the adults of African societies who are transformed, in the eyes of those they are appointed to initiate, into supernatural beings by the expedient of a coating of whitish earth? And Jeanmaire believed he could conclude that the Orphics' myth simply transposed the dangers undergone by an adolescent during the rites of passage.[65] In doing so he neglected all the sacrificial connotations of the narrative. He likewise forgot that in Orphism the myth was inseparable from an anthropogony: it concerns the origin of man, not the education of the young Dionysos, who does not accede to the adult status of his supposed initiators, the Titans.[66] They are instead his assassins, whom the thunderbolt reduces to ashes whence springs the human race.[67] Neither does the detail of the gypsum contradict the meaning of the entire narrative.

Technically speaking, gypsum (*gúpsos*) is a plaster the Greeks rarely used as paint. In Minoan structures gypsum was most often used in slabs for thresholds, orthostats, and the bases of columns.[68] But this white stone is often associated, not to say confused, with quicklime, a substance obtained by heating marble and limestone in kilns. It happens that the specific word for quicklime in Greek is *títanos*, or "titan," which signifies the whitish dust, the kind of white ash produced by the firing of any kind of limestone.[69] Consequently, the murderers masked in gypsum—are they not in fact hidden by that which best reveals their identity?[70] The supposition is apparently supported by certain traditions concerning the existence of the first autochthonous people, sometimes called *Titán*, sometimes *Titénios* or *Títakos*.[71] According to Philokhoros, the best known of the Atthidographers, the Attic land was once called the "Titan land," *Titanìs gê*, a name given it by a certain *Titénios* who lived in the vicinity of Marathon: he was older than the Titans, and the only one, it was said, not to fight against the gods.[72] *Titénios* was doubtless one of the princes whose autochthony went back to the time before

Kekrops, as Istros seems to admit in his collection of stories about Attica.[73] And among the people of Sikyon the first earth dweller was likewise named *Titán*: he was the brother of the Sun and had given his name to the whole region, which was thenceforth called *Titanê*.[74]

All these traditions concern beings born from the earth and, in particular, formed from the earthly element mixed with the fire designated by their name *títanos*, quicklime. The Titan of Sikyon has the sun as his brother. This mixture of fire and earth is also a physical fact recorded in Aristotle's *Meteorologica*: "Bodies formed from earth are mostly hot as a consequence of the heat that produced them, for example quicklime and ash, *títanos kaì téphra*."[75] But Eustathios's commentaries on Homeric epic preserve a much more precise relationship between the first autochthonous creatures called Titans and the enemies of Dionysos covered with whitish dust. In the margins of the Iliadic lines that invoke the white peaks of Titanos,[76] the learned archbishop of Thessalonika reminds us that *títanos* is the technical term for quicklime and that this name comes directly from the punishment suffered in the myth of the Titans when they were reduced to ashes by Zeus's fire and mingled with the white dust produced by the firing of lime and marble.[77]

So the Titans of the Orphic myth are not to be confused with the adversaries of Zeus in Hesiod's *Theogony*. In the tale of Dionysos's murder, the powers that take part are at once gods and men: gods, because like them they precede the human race, but also men because the affinities of these personages with a certain earthly substance, quicklime, qualify them to play the role of ancestors to a race as deeply rooted in the soil as the "eaters of bread." In a whole series of traditions as marginal as the preceding ones, the human race, which Hesiod's *Theogony* represents enjoying the gods' table before being separated from it by the sacrificial division, instead makes its appearance either by emerging from the entrails of the earth[78] or by being brought to life under the fingers of a demiurge who fashioned it by mixing earth and water, according to some,[79] or earth and fire, as others claim.[80] Orphic belief resuscitates on purpose a whole section of narratives circulating at the margins of official tradition, and it reorganizes them in a theological myth whose composite character the Greeks themselves recognized by condemning it as a forgery of Onomakritos, who supposedly—according to Pausanias—adapted the holy rites of Dionysos, borrowed the name Titans from Homer, and thought up the idea of making these characters the source of Dionysos's woes.[81] Though the analysis I have just made leads to other sources than the text of Homer, it also seems to show more clearly the Orphic origin of this myth, since the Titans, Dionysos's murderers, usher in immediately the centerpiece of Orphic theology, its anthropogony: from their ashes men will be born. Furthermore, the Titans coated in gypsum are not just the ancestors of the human race; they also, by virtue of their status as primordial beings and their half-earthly, half-fiery nature, dramatically prefigure the first human beings. By slaughtering Dionysos, by cutting up his limbs before cooking them in the cauldron and on the spit, and finally by feeding on him,[82] the Titans are already men who perform the ritual acts of the blood sacrifice and savagely assassinate the innocent victims they will feast upon.

This is the moment for a more systematic confrontation of the Titans of Orphism with Hesiod's Prometheus. The very place in which the first sacrifice of the ox transpires invites questions as to the actors and their intentions. Prometheus officiates at Mekone, a place in the region of Sikyon to which an autochthonous Titan once gave his name. Like the murderers of Dionysos, Hesiod's Prometheus is of Titanic nature. But Prometheus is above all the herald of the gods. He brings about the sacrifice and effects the split in his function as mediator. But he is not the demiurge molding human clay who is spoken of in certain traditions set aside by Hesiod. The difference is crucial, for it relates to the orientation of the first ritual sacrifice. As for other, more self-evident differences—the fire and the victim—they are minor. One of them, the fire, is worse than illusory, since the Titans' double cooking process makes it necessary, as we have seen, to presuppose the procedures of the official ritual whose initiator remains Prometheus.[83] This is so despite Hesiod's silence about the two ways of cooking. As for the victim, the choice of the ox in one and a young child in the other is certainly not insignificant, but the difference only takes us back to what is essential, namely, the intention of the sacrifice. In both cases a justification of man's position in the world is at stake. The Prometheus myth told by Hesiod is centered on the links of communication the sacrifice establishes between men and gods across the gulf that separates the two. Prometheus serves as their intermediary by making up the difference. From the same animal victim, which has been chosen because it is close to man yet is not confused with him, each will receive the portion that confirms his difference. On the contrary, in the Orphic myth, the Titans have no mediating function. All their behavior casts them on the side of men, and the gods are only present by the intermediary of the child playing the role of victim. This time the myth insists not on the gap that can be bridged but rather on the perdition and misery of the human race, here condemned to rise from the remains of a crime, a crime renewed in confused ignorance by the daily acts of those who believe they are giving thanks to the gods by devouring the flesh of slaughtered victims.

Thus, the overall staging of the Orphic tale tends to show that the alimentary blood sacrifice is, in the exemplary time of myth, a crime, an act of cannibalism, an allelophagic feast. Its first perpetrators were the Titans, who were also the first creatures sprung from the earth, and they hurled themselves with abandon into the murder of a child whose limbs, carefully cooked, they chewed up and swallowed. The horrible treatment inflicted in the myth on Dionysos assumes the appearance of the culinary process specified for the sacrificial ritual, with but one essential difference: the inversion of roasting and boiling, whose meaning and purpose we can now understand still more clearly. By adopting the scheme "boiling followed by roasting," the Orphics are actually striving to deny the process that on a culinary level makes the sacrifice a positive act with "progressive" connotations. To go from boiling to roasting or to roast boiled meat is to invert the sacrifice from within while respecting its apparent form, to destroy it from within after having condemned it from without. The sacrifice is an evil, and nothing can alter its fatal direction.

The murder of Dionysos by the Titans is a direct illustration of Orpheus's major teaching: "to abstain from murders, phónoi," a two-part exhortation to refrain from eating meat and to put an end to the assassination of human beings. Through this myth Orpheus taught men they must utterly refuse to engage in the blood sacrifice. This ritual, far from establishing relations with divinity, reproduces in disguise a crime in which mankind will never cease participating until it has realized once and for all its Titanic descent and undertaken by means of the so-called Orphic way of life to purify the divine element shut up inside it by the voracity of those who lately slew the young Dionysos.

Modern exegetes of Orphism, when they do not simply deny all credence to the myth of the Titans,[84] have often objected that the slaughter of Dionysos was not necessarily linked to the Titanic origin of the human race.[85] The philological record seems indeed to invite skepticism. It is true that the Plutarchean treatise On the Eating of Meats is the first to attest an explicit homology between the Titans devouring Dionysos and carnivorous mankind.[86] It is not less true that no one before Dio Chrysostom tells of the birth of men from the Titans' blood.[87] There is a lapse of six centuries between these sources from the first century A.D. and Onomakritos, a contemporary to the beginnings of Orphism. A philologist does not need more to feel properly uneasy. But the analysis we have just made allows a response to these strictly philological objections. From the time of its oldest version in the Aristotelian *Problems*, the myth told in the Orphic poem entitled *Teletē* ("Rite of Initiation") contains certain details whose explanation demands recourse to eating rules considered the essence of the Orphic way of life by the oldest traditions, those of Aristophanes and Plato. We have seen that the treatment inflicted on Dionysos by the Titans is stripped of meaning unless it is referred to the ritual of the alimentary blood sacrifice. Yet this was the essential problem debated among the sects: to eat or not to eat meat. Among Pythagoreans or in the city, particularly in the narratives about the ritual of the Murder of the Ox (*Bouphónia*), the only issue is how to fix man's position in relation to beasts and gods, which is why the central myth of the Orphics so naturally deploys an anthropogony.

Up to now our analysis has been restricted to deciphering the story of Dionysos and the Titans in terms of a sacrificial model whose features it seems to adopt by inverting them. It has been necessary to investigate en route the identity of Dionysos's murderers in order to understand the theological meaning of the myth, but we have apparently neglected to consider the victim and the problems he poses. For if the Titans were chosen by the Orphics for the reasons we have sought to discover, the choice of Dionysos as the victim of this anthropophagic feast is doubtless not gratuitous either. In fact, his inclusion in the discourse the Orphics develop on the denial of the blood sacrifice poses at least two questions. The first is a problem of theological architecture within Orphism itself: what is the place of Dionysos as a divine power in the theogonic thought of the Orphics? As for the second, it concerns the relations between Orphism and the Dionysiac movement. Is it by chance that the Orphic theology retained in place of the ox or the animal victim a god whose followers are pleased to invoke as an "eater of raw flesh"?

Of these two problems, the first flows directly from a detail of the narrative that our analysis has yet to acknowledge, though it is a particular of a culinary kind. Certain versions of the death of Dionysos stress neither the crime nor the punishment of those responsible but instead the fate of the victim. Once Dionysos has been boiled and roasted, he is carefully divided among the table mates; nevertheless, at the conclusion of the banquet he will be reborn. Aside from two versions that speak of the gathering of Dionysos's limbs—in one case by Zeus, in the other by Demeter-Rhea[88]—the other versions closely associate the rebirth of the victim with the special treatment of one of his organs. "Of the child's limbs," says one, "they made seven shares, but they abandoned just one piece, the heart endowed with intelligence."[89] Whether it is simply absent from the division or saved by Athena who steals it, Dionysos's heart is the only part of the victim the Titans do not consume. Its privileged status is confirmed by a whole ritual tradition forbidding that the heart be eaten.

Normally this organ is one of the viscera, the *splánkhna*, served on skewers at the beginning of the sacrifice. But several cult regulations proscribe the victim's heart as food. In an inscription that details the ceremonies celebrated at the prytany of Ephesos by the College of Kouretes,[90] it is provided that the prytany light the fire, burn spices on all the altars, and provide the victims at its own expense.[91] The number of them totals 365, of which 190 must be offered *kardiourgoúmena* and *ekmērizómena*: the thighs are to be cut out for consumption in honor of the gods, while the hearts will be extracted from the victims,[92] i.e., removed from consumption. In this context the specific reasons for the prohibition escape us, but other documents allow us to retrieve them by disclosing the relationship between this tabu and others of the same kind. Thus, a sacred law from Rhodes dated to the first century A.D. stipulates that whoever wishes to penetrate a sanctuary (of Asklepios or Sarapis?[93]) in a pure state, *hágnos*, must respect a triple prohibition: neither sexual pleasure, nor broad beans, nor heart.[94] This regimen is less severe, however, than the one imposed on the priests of Zeus and Athena in their sanctuary on Mount Kynthos on the island of Delos: neither women nor meat diet for as long, apparently, as their priesthood lasted.[95] At Rhodes the continence is temporary and the abstinence partial: just broad beans and heart. The latter two prohibitions crop up again among the Pythagoreans, though they are not their exclusive property among mystics. The disciples of Orpheus and the initiates of Eleusis do not eat broad beans either.[96] But Pythagoreanism provides the context that poses in explicit terms the best justification for this two-fold prohibition fundamental for all Greek mysticism.

The broad bean is indeed the most marked source of generation in the plant world, to the extent that it appears as a mixture of blood and genitals in the fantasies of Pythagoreans. But the prohibition against tasting it only restates in more urgent terms the hackneyed tabu against eating meat or spilling the blood of a living being.[97] We also know that in this same mystic domain the denial of the blood sacrifice can take the weakened form of a prohibition against eating certain parts of the victim, the most frequently named being the heart and the brain. "Don't chew the heart," says a Pythagorean precept; "don't eat the brain" adds another often joined to it.[98] The heart and the brain are actually the source of generation, *geneseōs arkhē*, in the living being. This phrase—"principle of birth"—is used by an interlocutor in Plutarch's *Table Questions* who thinks it necessary to search in Orphic or Pythagorean doctrines to explain why one of the guests has just refused an invitation to eat an egg.[99] Like the heart or the brain, which present great affinities to sperm in Pythagorean eyes,[100] the egg cannot be eaten because it is the living creature *par excellence*, as is attested by the Orphic representations of the primordial Egg from which some say Phanes-Eros sprang, or whose broken shell, according to others, will give shape to earth and sky.[101] Thus egg, heart, and broad bean occur together in one and the same list of prohibitions that a cultic regulation from Smyrna dated to the second century A.D. enumerates for the initiates of Dionysos Bromios:[102] don't approach the altars in black clothing, don't strike victims that cannot be sacrificed, don't serve eggs in the banquets honoring Dionysos, burn the heart (of the victim) on the altars,[103] and abstain from mint,[104] which accompanies the cursed race of broad beans. But the horror provoked by these leguminous plants in this Dionysiac context receives a novel justification that the regulation suggests be told to the mysteries' initiates: broad beans were born from Titans, the murderers of Dionysos.[105] If this document composed in dactylic hexameters is, as A. D. Nock suggested, a product of the oracular activity in Asia Minor during the second century A.D.,[106] its originality does not lie in the way it combines Orphic precepts with Pythagorean tabus while giving them a certain Dionysiac flavor. It lies instead in the way it overdetermines parallel motifs of different origin, for example, by inventing a Titanic origin for broad beans. These vegetables are not only the nocturnal, bloody doubles of the human plant, whose consumption is tantamount to cannibalism; broad beans are themselves born from the primordial beings who were promoted to the rank of ancestors of carnivorous humanity by their anthropophagic behavior. The cannibalism with which broad beans were passively marked has been reinforced by the allelophagy actively attested by the Titans. The horror that the Titans in their turn inspire renders still more intense the repulsion inspired by broad beans. But, contrastingly, the same context of mystical prescriptions happens to prove that there is no directly dependent relationship between the prohibition against eating heart as formulated by the rituals and the conduct adopted by the Titans in the Orphic myth. The heart is here forbidden for the same reason as the egg: because it is the principle of life, *arkhē*, in its double sense of that which begins and that which rules.[107]

Greek doctors and physicists, each after their fashion, systematized this representation of the heart.[108] Philo of Alexandria provides a banal version of it that is an effective summary: "It is known that,

according to the best doctors and physicists, the heart takes shape
before the rest of the body, just like the foundation of a house or the
keel of a ship. And they say that the heart still palpitates after death
so that it disappears last just as it appears first."[109] It is first and
last like the goddess of the hearth, Hestia, to whom the Aristotelian
treatise *On the Parts of Animals* compares it: "The heart is necessary
because it is the principle of heat, and a kind of hearth, *hestía*, is
needed to keep the flame of nature. This hearth must be well kept,
since it is like the citadel of the body, its *akrópolis*."[110] For all living
creatures that have blood, the heart is the first organ to be formed;
it is the first product of the bloody humor whose coagulation gives
the viscera their shape. As the principle of blood that quickens the
living, the heart is also its first receptacle. It is the part of the body
that begins to move, like a little animal, before all the others.[111] The
heart, which is born first, is also the organ occupying the first position, the middle, which is a privileged position because it is unique
and because it can be reached from all points equally, or almost
equally. In the same work that develops this representation at length,
Aristotle insists on the heart's central position. It is lodged in the
middle of the thoracic cage, even if it must be conceded that in the
human body it sits slightly to the left in order to compensate for
the coolness of a part that is less privileged since it is other than the
right. The only place the heart can occupy is the center, the most
valued point in space, the source of the *arkhé* from which commands
issue and everything begins.[112] Philolaos exploits this same model on
a cosmogonic plane with the theory of a central fire, *hestía:* the earth
takes shape under the behest of the middle and beginning from it;
it grows equally up and down.[113]

To be sure, no evidence states explicitly that Orphism shares this
conception of the heart, but Aristotle is our guarantee that the "so-
called" verses of Orpheus contained a theory of the development of
the living being according to which the parts of the body formed one
after another, as one plaits a net.[114] The heart was certainly not the
last-born. The story of the Titans leads one to believe quite the opposite. If the heart is the only part of Dionysos that escapes destruction, that is because the preservation of this organ, the most vital in
every animate being, allows the god to be reborn even after he has
been eaten. To repeat the traditional discourse referred to by Aristotle: the heart is only the last because it is first of all the first.

Such is precisely Dionysos's status in the Orphic theology: the last
member in a series of which he is also originator and principle. In the
system developed by the *Rhapsodic Theogony*, six divine generations
succeed one another. Phanes, also called Mêtis, is the first to spring up

in a blinding flash of light. He then yields the scepter to the sovereignty of Night. Ouranos, followed by Kronos, succeeds her, and Zeus is the fifth sovereign whose power is established with the advice of Night and the complicity of Phanes-Mêtis, whom the new king of the gods devours. Zeus finally entrusts the royal power to his son born from union with Persephone: Dionysos will be at once the first and the last sovereign, since the sixth generation, which concludes the series, also makes a return to the beginning.[115] Dionysos is only another name for Phanes.[116] By Zeus's mediation, the original First-Born is identified with the last-born, who is the new young king of the world and the gods.[117] His rebirth closes the circle of divine generations just as his slaughter opens the cycle of births and generation for the human race.

The central role assumed in Orphism by Dionysos on the cosmogonic, theogonic, and anthropogonic levels poses with new urgency the question of the relations between this form of religious thought and the mystic movement developing in Dionysiac religion at the same period with entirely different results. The question is even more pertinent since it can be articulated on the sacrificial plane with respect to actual procedures and the status of the victim as well. Though some modern scholars have been seduced by the altogether factitious resemblance between the omophagic ritual of tearing to pieces the victim in Dionysiac cult and the slaughter of Dionysos ritually sacrificed by the Titans, others, who are more circumspect but still convinced that Orphism was a kind of protest movement within Dionysiac religion, were content to observe that the Orphic myth disguised the Dionysiac sacrament of omophagy as a crime.[118] To do so was to recognize the importance of sacrifice for the two mystic movements. In fact, omophagy is to Dionysiac religion what the refusal to eat meat is to Orphism. Eating raw food is also a way of rejecting the blood sacrifice and the system of values wedded to it. In hacking up the body of a wild, not a domestic, animal that has been captured after a violent chase, in chewing up its raw meat instead of eating only certain pieces that have been roasted or boiled, the person possessed by Dionysos annihilates the barriers erected by the politicoreligious system between gods, beasts, and men. Swept away by the "eater of raw flesh" and his wild hunt, the devotees of Dionysos ōmēstés cease to be tranquil diners on the flesh of an animal that has been cooked by the rules. They become savage themselves and behave like ferocious beasts. They escape the human condition by way of bestiality, taking the lower route among the animals, while Orphism proposes the same escape on the divine side, taking the upper route by refusing the meat diet that spills the blood of living beings and eating only perfectly pure food. Dionysiac omophagy is the homologue of Orphic vegetarianism. Both movements aspire to the same goals, but by different procedures, which, nevertheless, travel complementary routes.[119]

Even so, the inverse relationship between their chosen means—in one case from above, from below in the other—does not imply that the Orphic myth should feature an inverted image of the Dionysiac ritual. The relations are more complex, for the complementarity in routes traveled does not happen without competition on the same terrain of the refusal to practice the blood sacrifice demanded by the city-state. In the central Orphic myth there is a certain distance taken with respect to Dionysiac cult. The all-powerful eater of raw flesh becomes the pathetic victim of a band of cannibals in the story of the Titans. By a kind of irony, Dionysos is eaten after having been cooked in the cauldron and on the spit just like any animal victim in the type of blood sacrifice Orphism rejects with as much conviction as Dionysiac religion. It is as if in a single mythical narrative the disciples of Orpheus had wished to condemn the carnivorous life of the city and at the same time to suggest discreetly, through the death of Dionysos, the inadequacy of a diet in competition with the one that Orpheus wished to impose. Inadequacy or perhaps danger, since omophagy often borders on allelophagy. The frenzy engendered by the consumption of a live animal ripped apart can lead to excessive behavior, as the legendary Bassaroi attest: in the delirium of human sacrifice, after having devoured their victims, they hurled themselves upon each other, tearing off strips of flesh with their bare teeth.[120] Again, there are the daughters of Minyas, who were smitten with the desire to taste human flesh and capable of ripping to pieces their own child. They are the ones who justify the ritual scenario of the Agrionia, in which the priest of Dionysos, armed with a sword and chasing the descendants of these banshees, had the right to slaughter whichever one he could catch.[121] In several parts of the Greek world, Dionysos's savagery is no less disturbing than the Titans' violence: he is the god who craves human sacrifice, who eats human flesh.[122] The disciples of Orpheus had serious reasons for thinking that Dionysiac religion sometimes bore a strange resemblance to the murderous madness they denounced in the meat diet and the sacrifice in use among those who remained prisoners of the city.

Furthermore, there is a profound antagonism between Orphism and the Dionysiac movement on the level of the sacrifice. For neither the murderous violence authorized by his madness nor the cannibalistic frenzy he inspires in some of his devotees are disavowed by the sacrificial practice that is encouraged by the disciplined form of Dionysiac worship nicely assimilated into the "political" religion. The slaughter of a he-goat in honor of Dionysos is as banal an act in the city-state as the offering of a sow on Demeter's altar. Some cities even seem to have accepted a weakened form of the raw diet. At Miletus in the third century B.C., the priestess of Dionysos Bacchios performed on behalf of the city the ritual gesture called ōmophágion embállein.[123] She places in the holy basket a "mouthful of raw meat," which is not the substitute for some regular animal sacrifice but the perfectly discreet vestige of the great hunts for fresh flesh that Dionysos loves to lead over hill and dale. As close and familiar as Dionysos can make himself, he is always identifiable with bloodshed, as opposed to the pure life laid claim to by Orphism.

Such a sharp split on the essential level of sacrificial behavior would lead one to believe that the disciples of Orpheus fabricated from whole cloth an anti-Dionysos for ideological uses in their theological discourse. But if there is forgery involved or rather, as we believe, resemanticization, it is not at the expense of such a simple, univocal Dionysos. Dionysos is more complex and polymorphous than any other divinity in the pantheon—by his rare prestige as a magician as much as by his affinity for displaying or manifesting in the *beyond*. His *beyond* with respect to the human condition between gods and beasts does not only take the form of the state of cruel bestiality omophagy imposes. For the very same Dionysiac indistinctness between men and beasts likewise leads to the disappearance of any distance between men and gods. The extreme savagery the god's possession demands actually takes the form of a golden age made present by the absence of any difference between man, god, and animal. This golden age of Dionysos is attested in various places, above all in the tales of his childhood. The eater of men who transforms women into ferocious beasts grew up in the land of cinnamon, in perfumed Ethiopia, where Herodotus's account localizes "Holy Nysa."[124] A hundred perfumes emanate, ewes are covered with wool, springs gush up, and distant birds bring boughs of cinnamon. Such is the birth of the divine child told by Dionysius the Periegete in the second century A.D.[125] Seven centuries before the Bacchae of Euripides' tragedy pass without transition from a paradisiacal state to the wild hunt.[126] At the call of the thyrsos, a fresh spring flows from a rock, wine gushes from the earth, and fingers that scratch the soil are moistened with milk, while honey drips from ivy leaves. The Mainads give suck to young wolves and snakes come to lick their cheeks. Suddenly the Bacchic race begins, "the mountain with its savage creatures goes mad." The bacchants alight on a herd of cattle, tear up their coats, rip their flesh, twist off their limbs. Then the band, "like a flock of birds taking flight," crosses the wheatfields of Demeter and pounces on the villages at the foot of Kithairon. Everything is laid waste. The Mainads carry off the children and along with them bronze, iron, and fire—all the tools of civilized life, including cauldrons, spits, meat-hooks, fireplaces,[127] as if the triumph of Dionysos and his savage paradise meant the negation of the possibility and the instruments of sacrifice. Elsewhere, too, there is the same imperceptible slippage from one pole to the other in an indifferent alternation of extremes. For the daughters of Minyas who choose to stay at home rather than go and join the thiasos in the mountains, the uprights of their looms begin to flow with nectar and milk. But this time Dionysos only reveals the dazzling products of his golden age after showing the face of his bestiality—bull, lion, and leopard in succession. And the motion is reversed in the following episode. The nectar that sabotages the technical object changes into a body ripped to pieces by the obstinate weavers into whom Dionysos breathes an irresistible yearning for the taste of human flesh.[128]

Of the two orientations always available in the Dionysiac movement, Orphism wished to retain only the one that confirmed its decision to short-circuit the politicoreligious system by golden age practices. Consequently, it had to put an end to the circular interchange between the two poles linked by Dionysiac strategy. It had to impose on Dionysos the violence of a split necessitated by its abhorrence of bloodshed as well as the fundamental choice of Orphism in favor of the pure man, by which is meant masculine purity. The enactment of Dionysos's murder by the Titans was designed to exorcise the cannibalistic frenzy that haunted the sacrificial act. The pathetic victim of a monstrous crime, Dionysos found himself abducted from his murderous orgies now abandoned to Bacchants and Mainads. He was reborn child sovereign of the gods, charged with inaugurating the reign of remade unity in the absence of wild cries and bodies ripped to pieces. In this way, Orphism modified a part of Dionysiac religion, and it doubtless contributed to its tendency to become a salvation religion.[129] Furthermore, it succeeded in bridging the theological gap[130] of a movement that seems to have been richer in initiation rites than in exegetic discourse,[131] as opposed to Orphism, each of whose initiation rites, or teletai, took the form of a narrative of theological character, a text written in the margins and recesses of another, to such an extent that its voice swelled to that "tumult of books" that gave Plato a laugh.[132]

But to mark Dionysos with their choice the Orphics had to stigmatize his savagery in such a way that it be openly censured and denounced without qualification. The result can be seen in the myth of Orpheus's death recounted in a tragedy of Aeschylus from his tetralogy on the Lykourgos story: the *Bassarai*, or Bacchants.[133] Every morning, Orpheus climbs to the summit of Mount Pangaion to greet the appearance of the sun, whom he identifies with the god Apollo.[134] And Dionysos, who rules this landscape, avenges the scorn he is held in by Orpheus and entrusts him to the Mainads' rage. So Apollo's fanatic devotee is torn to pieces by Dionysos's followers. The Apollonian mysticism that the disciples of Orpheus cherished along with the Pythagoreans here appears in radical opposition to the behavior Dionysos inspires. "The great priest of Thrace in his long, white surplice"[135] meets a band of women hungry for human flesh. These bacchants are called Bassarai, the same name borne by the sane priestesses who lead the thiasos at Ephesos or Torre Nova,[136] but also homonyms of those worshipers of Dionysos whose dangerous madness Porphyry tells us drove them to devour each other, intoxicated with the pleasure of having tasted the flesh of the human victims they sacrificed.[137] The split between Dionysos and Apollo here functions inside another, more profound, split that Orphism made its own: between woman and man, between the impure bestiality of one and the pure spirituality promised the other. Orphism exiles the savage violence of Dionysos into the animal world of woman, who is thus, by her very nature, excluded from the Orphic rule.[138]

Some readers of myths, seduced by a compelling taste for the concrete, would surely like to see Orpheus's misadventure as an episode bewailing the disputes that set Dionysos's partisans against the harmless disciples of Orpheus.[139] But the history of social turmoils is nothing more than the chronicle of cataclysms directly transcribed in the discourse of myth. The confrontation between a wild Dionysos and an Apollonian Orpheus only takes on meaning when it is itself confronted with the strictly complementary relationship between the Delphic Apollo and the Dionysos whom the Titans devour. If one version finds Apollo instead of Demeter or Rhea gathering the remnants of the slaughtered child,[140] and if in several versions Dionysos is welcomed at Delphi in the most Apollonian sanctuary,[141] it is because Orphic theology makes a strict distinction between the golden age Dionysos, sovereign of the refound unity, and the god of omophagy, prince of bestiality.

One can still ask if the split inflicted by Orphism on the body of Dionysiac religion is not menaced by the very thing that makes it possible, that is, by Dionysos's inveterate oscillation between the twin poles of savagery and paradise regained. By conferring pride of place on a divine power whose royalty at the beginnings of the world is based on his privileged ability to reunite in himself the most diverse forms and elements, Orphism was unmistakably drawn into the orbit of the Dionysiac phenomenon and caught up in the whirl of metamorphoses that Dionysos's constitutive, ever-renewed madness indulges in throughout the course of history.

The myth, then, amounts to theologian's discourse, since it concerns itself with determining the position of a divine power in a system of thought that pivots on the problem of the one and the many. Dionysos has been run to earth, but analysis there proves to be interminable. Drawing an arbitrary and provisional line, we suspend the game and revert to a review of the analytic process. Our first step has not been to challenge on their own ground the interpreters convinced that the myth must reflect the scenario of a Dionysiac ritual or that the consumption of a divine child could be reduced by a comparatist's jiggling to the "natural" representation of a god whose essense is to die and be reborn. At the outset there is a series of strange questions, beginning with the culinary and sacrificial paradox embedded in the center of the Orphic myth. But these questions could not even be formulated unless other narratives, ritual practices, and different mythical traditions had been invoked, confronted, and set into relation with another beforehand. Bringing to light the unusual character of a story about cannibalism in a sect obsessed with the horror of bloodshed, or remarking a whimsical use of culinary practice—these are part of the attempt to interpret the organized semantic setting without which the decipherment of the myth can neither commence nor take direction. The strangeness of the Orphic narrative becomes clear as its matter is confronted with sacrificial procedures, with the crucial relationship between spit and cauldron, and with the whole set of significations the Greeks gave to roasting and boiling. Likewise, by specifying the values of the gypsum with which the Titans are smeared, the actors of the myth reveal themselves as primordial men sprung from whitish earth and associated with quicklime.

The story of the death of Dionysos is referred, by its most pertinent features, to an ensemble of chiefly mythical representations organized around dietary customs, cooking processes, the blood sacrifice, and by way of them, the human condition as defined in its relation to beasts and in its relation to gods. Thus, the murder of Dionysos by the Titans belongs to a group of myths comprising the story of Prometheus, the representations of Dionysiac omophagy, and the Pythagorean speculations about the death of the plow ox. But the group could be enlarged in two directions to include, first, the various stories that the city-state developed around the *Bouphónia* ritual and that are themselves a subgroup of the myths centered on the first animal bloodshed; and secondly, the totality of mythical and ritual traditions that form the history of Dionysos and constitute the fragment of a mythology whose interpretation is becoming urgent if only in the immediate perspective of an extended confrontation between the Dionysiac and Orphic movements.

This myth, which is central in Orphic thought, does not entrust its meaning to the apparent, more or less explicit, sense of the surface relations that bind its characters together: the Titans, Dionysos, men, gods. On the contrary, it has been necessary to shun their pressing message in order to recognize the semantic relations that each term of the Orphic myth entertains with other mythical narratives or various ritual data. This is the only way to appreciate the work of reorganization invested in a discourse conceived and contrived by theologians. Moreover, they were marginal theologians whose discourse advertised their opposition to politicoreligious thought and their rupture with Hesiodic myths and the city-state system, who selected at various levels, from both marginal and common traditions, the terms and relations they would reinterpret and combine into a deliberately learned discourse that could only proceed circuitously and by way of studied connotations.

Notes

Chapter 1

These remarks were the subject of a presentation at the colloquium on Greek mythology organized by the International Center of Semiotics and Linguistics, Urbino, 7–12 May 1973. The notes, minimal under the circumstances, were expanded in the bibliography developed in two studies by Claude Calame: "Philologie et anthropologie structurale. A propos d'un livre récent d'Angelo Brelich," *Quaderni Urbinati di Cultura Classica* 11 (1971):7–47; and "Mythologiques de G. S. Kirk. Structures et fonctions du mythe," ibid. 14 (1972):117–35. The review *Critique* (332 [January 1975]:3–24) has already set these pages before its readers.

1. Claude Lévi-Strauss. *Mythologiques II. Du Miel aux Cendres* (Paris, 1966), p. 407 [*From Honey to Ashes: Introduction to a Science of Mythology*, trans. J. and D. Weightman, vol. 2 (New York, 1973), p. 473].

2. Claude Lévi-Strauss, "La Structure des mythes" (1955), reprinted in his *Anthropologie structurale* (Paris, 1958), pp. 227–55 [*Structural Anthropology*, trans. C. Jacobson and B. G. Schoepf (New York, 1963), pp. 202–28].

3. Ibid., p. 238, n. 1 [English trans., p. 209].

4. Claude Lévi-Strauss, "La Geste d'Asdiwal," *Annuaire de l'Ecole pratique des Hautes Etudes, Sciences religieuses* (1958–59) (Paris, 1958), pp. 3–43, reprinted in his *Anthropologie structurale deux* (Paris, 1973), pp. 175–233 [*Structural Anthropology II*, trans. Monique Layton (New York, 1976), pp. 146–97, with a postscript].

5. Edmund Leach, *Genesis as Myth and Other Essays* (London, 1960), as well as, among others, *Claude Lévi-Strauss* (New York, 1970); G. S. Kirk, *Myth: Its Meaning and Functions in Ancient and Other Cultures* (Cambridge, 1970), pp. 48 ff.; and M. Meslin, *Pour une science des religions* (Paris, 1973), pp. 223–24.

6. Cf. Georges Dumézil, *Les Dieux des Indo-Européens* (Paris, 1952), pp. 80 ff.

7. Cf. Georges Dumézil, *Mythe et épopée* (Paris, 1973), 3:16.

8. Cf. Paul Zumthor, *Essais de poétique médiévale* (Paris, 1972), pp. 75 ff.

9. Pp. 565–83, expanding on *Myth*, a book of greater scope.

10. Cf. below, 68–94.

11. Pierre Smith and Dan Sperber, "Mythologiques de Georges Dumézil," *Annales E.S.C.*, 26 (1971): 583–85.

12. Cf. Marcel Detienne, *Les Jardins d'Adonis: La mythologie des aromates en Grèce* (Paris, 1972) [*The Gardens of Adonis: Spices in Greek Mythology*, trans. Janet Lloyd (Sussex, 1977)]; idem, "Orphée au miel," in *Faire de l'histoire*, ed. Jacques Le Goff and Pierre Nora (Paris, 1974), 3:56–75.

13. Cf. Dan Sperber, "Le Structuralisme en anthropologie," in *Qu'est-ce que le structuralisme?* (Paris, 1973), postface; idem, *Le Symbolisme en général* (Paris, 1974), "III. La Signification absente."

14. Hans Schwabl, "Hesiods Theogonie. Eine unitarische Analyse," *Sitz. Oesterr. Akad. d. Wissenschaften, Phil. hist. Klasse*, B. 250, 5, Vienna, 1966.

15. Jean-Pierre Vernant and Pierre Vidal-Naquet, *Mythe et tragédie en Grèce ancienne* (Paris, 1972); Pierre Vidal-Naquet, "Oedipe à Athènes," preface to

Tragédies de Sophocle (Paris, 1973), pp. 9–37; Nicole Loraux, "L'Interférence tragique," *Critique* 317 (October 1973): 908–25.

16. Claude Lévi-Strauss, *L'Origine des manières de table* (Paris, 1968), pp. 186–89.

17. Paul Faure, "Aux Sources de la légende des Danaïdes," *Revue des Etudes grecques* 82 (1969): 186–89.

18. Claude Lévi-Strauss, "Structuralism and Ecology," *Barnard Alumnae* 1, (1972): 6–14.

19. Jean-Pierre Vernant, "Le Mythe hésiodique des races" (1960), in *Mythe et pensée chez les Grecs*, 3rd ed. (Paris, 1971), 1:38–41; Marcel Detienne, *Crise agraire et attitude religieuse chez Hésiode*, Collection Latomus, 68 (Brussels, 1963).

20. Jean Pouillon, "Présentation," *Problèmes de structuralisme*, *Les Temps modernes*, no. 246 (November 1966), p. 784; see also idem, *Fétiches sans fétichisme* (Paris, 1975), p. 23.

21. Cf. below, 53–67.

22. Pouillon, "Présentation," p. 783; see also idem, *Fétiches*, p. 23.

23. Brian Vickers, *Towards Greek Tragedy* (London, 1973).

Chapter 2

1. That "rupture of history" spoken of by Pierre Nora—historians and sociologists of the Greek world not only live it but contribute to it, as much by their practices as in the questioning of the procedures they are principal heirs to, having continuously followed the course history assigned them since the sixteenth century.

2. Ephoros in *FGrH* 70 fr. 110. Cf. G. Nenci, "Il motivo del' autopsia nella storiografia greca," *Studi classici e orientali* 3 (1966): 35–38; and G. Schepens, "Ephore sur la valeur de l'autopsie," *Ancient Society* 1 (1970): 163–82.

3. Pierre Boyancé, "Comment on écrit l'histoire," *Revue des Etudes anciennes* 73 (1972): 157. For an epistemological reading of the same essay by Paul Veyne (Paris, 1971), see the work of Michel de Certeau, "Une épistémologie de transition: Paul Veyne," *Annales E.S.C.* 27 (1972): 1317–27.

4. L. Robert, "Epigraphie" (Les Epigraphies et l'épigraphie grecque et romaine) in *Histoire et ses méthodes*, under the direction of C. Samaran, *Encyclopédie de la Pléiade* (Paris, 1962), pp. 453–97, p. 462.

5. Ibid., p. 463. "A bulletin [of epigraphy!] that analyzes each year's new discoveries and publications is a true kaleidoscope."

6. Ibid., p. 461.

7. Ibid., p. 463.

8. Ibid., p. 462.

9. Ibid., pp. 461–63.

10. A place is assigned them under the rubric "Musées, pierres errantes, provenances" ["Museums, wandering stones, places of origin"] in the *Bulletin épigraphique* published since 1938 by J. and L. Robert. This journal, a mass of erudition, embraces the historical discourse of L. Robert. It is the place where *l'intrigue*, as Paul Veyne calls it, takes shape—pregnant singular of a plural that expands into an extraordinary social history cutting across the learned text. One speaker dominates this social history, thanks to whom one can retrieve some major systematic links between *knowledge* and *place* (Michel de Certau, *L'Ecriture de l'histoire* [Paris, 1975], pp. 63–120: "l'opération historiographique").

11. Robert, "Epigraphie," p. 475. Cf. Roland Barthes, "Le discours de l'histoire," *Social Science Information* 6 (1975): 65–75.

12. Marcel Detienne, *Les Jardins d'Adonis: La mythologie des aromates en Grèce* (Paris: Bibliothèque des Histoires, 1972). [*The Gardens of Adonis: Spices in Greek Mythology*, trans. Janet Lloyd (Sussex, 1977)].

13. Pierre Lévêque, "Un nouveau décryptage des mythes d'Adonis," *Revue des Etudes anciennes* 74 (1972): 180–85, p. 185.

14. Giulia Piccaluga, "Adonis e i profumi di un certo strutturalismo," *Maia* 26 (1974): 33–51. Be it literary, critical, or both at once, a text does not necessarily give itself away solely by its avowed discourse. In granting with such generosity the seal of his authority to the critique published by his review, the director of *Maia* does not seem to have suspected what was implicit in the discussion opened by Piccaluga. And it may be considered strange that such an informed literary critic, so bored with "exact exegesis" (cf. *Maia* 25 [1973]: 329), did not realize that the violent attacks addressed at *Jardins d'Adonis* were based on a major presupposition: that Adonis, son of Myrrha, belonged to the species of "discomfited hunters." Such is the thesis developed by Giulia Piccaluga under the title "Adonis. I Cacciatori falliti e l'avvento dell' agricoltura" at the colloquium on the analysis of Greek myths organized by the Centre international de Sémiotique et de Linguistique (Urbino, May 7–12, 1973) and published in *Il Mito greco*, ed. B. Gentili and G. Paioni (Rome, 1977), pp. 33–48. The debate will at present not concern method, and the reasons of opportunity conjured up in the preliminary note signed "ALP" are amply sufficient to explain the somnolence of the critical spirit. Was it not necessary to give an example in order to stem the structuralist tide and restore the authority of the philological-literary method which, as is well known, no longer needs to prove itself? I will make my response on the very terrain on which A. La Penna challenges structural analysis, condemning it for "geometricity" and preconceived schemes: an "exact exegesis" of the hunting activity that mythical tradition grants to Adonis, and around which, unknown to A. La Penna, Piccaluga is building the stronghold to protect her handsome mastery of interpretation. (A more detailed statement of my disagreements with P. Lévêque and G. Piccaluga can be found in the French version of *Dionysos Slain*, pp. 55–72.)

15. Lévêque, "Un nouveau décryptage," p. 182.

16. Piccaluga, "Adonis e i profumi," p. 43.

17. Cf. Roland Barthes, "L'effet de réel," *Communications* 11 (1968): 84–90.

18. This orientation, first adopted by Karl Meuli, is the basis of W. Burkert's research (*Homo Necans* [Berlin and New York, 1972]) on the fossilization in ritual of the murderous customs of that predatory hunter, Paleolithic man.

19. A. Brelich, *Paides e Parthenoi*, vol. 1 (Rome, 1969).

20. Pierre Vidal-Naquet, "Les Jeunes: Le cru, l'enfant grec, et le cuit," in *Faire de l'histoire*, ed. Jacques Le Goff and Pierre Nora (Paris, 1974), 3:137–68. The ensemble of traditions on the foundation of the she-bear ritual is examined by W. Sale, "The Temple Legends of the Arkteia," *Rheinisches Museum* 118 (1975): 265–84.

21. Hare and deer, on the cotylus cup of Amasis: Louvre A 479 (S. Karouzou, *The Amasis Painter* [Oxford, 1956], p. 37, no. 71, pl. 13); hare and fox, on the cup of Sokles: private collection (K. Schauenburg, "Erastes und Eromenos auf einer Schale des Sokles," *Archäologischer Anzeiger* 80 [1965]: 849–67, fig. 3).

22. Strabo, 10, 483C. Cf. H. Jeanmaire, *Couroi et Courètes* (Lille, 1939), pp. 450–55.

23. Euripides, *Hippolytus* 17.

24. Cf. J. Kambitsis, *Minyades kai Proitides*, vol. 1 (Jannina, 1975).
25. Antoninus Liberalis, *Metamorphoses* 21. The bear's desire is inspired in him by Aphrodite, who is outraged at the scorn borne her by Polyphonte.
26. Detienne, *Les Jardins d'Adonis*, p. 130 [English trans. p. 67]. The Athenian Adonis is a poor excuse for a hunter, and the secrets of his "colleague" Perdiccas will not save his reputation in this domain. From the first century A.D. on, another tradition is attested, which finds confirmation in the imagery of Sarcophagi depicting the *virtus* ["manliness"] of Adonis in the hunt, whether he is killed or triumphant (as he perhaps is on the mosaic of Daphne near Antioch, where, beneath the gaze of young "Magnanimity," *Megalopsychia*, Adonis stands killing a leaping boar). In the treatise on hunting he published at the moment Ovid left for exile, Grattius, who sees the sign of reason and the beneficence of Diana in the art of the hunt, wishes to show by two examples how deadly ignorance of the hunt can be. He refers to two heroes of old who were victims of their own mad bravura: Ancaeus and Adonis, both of whom fell beneath the blows of a boar (Oppian, *Cynegetica* 1, 24 ff.). It was not courage they lacked, but skill in hunting. Not until the fifth century A.D. in the *Dionysiaca* of Nonnos do we find, amid the adventures of Beroe, a story of Adonis the hunter marrying Aphrodite. Beroe, eponym of the town of Berytos, is a friendly huntress who also bears the name Amymone (some coins from Berytos at the time of Elagabalus show Beroe in the guise of Amymone surprised by Poseidon: Ch. Picard, "Le Poseidon lysippique de Bérytos," *Rev. arch.* 47 [January–June 1956]: 224–28). Dionysos and Poseidon contend for her hand; the sea god wins, delighted to grant his protection to the Phoenician city. In telling of the rivalry of the two suitors, Nonnos refers to a "recent" legend (41, 155–57) according to which Beroe was born of the love of Aphrodite and the Assyrian Adonis. A few verses below (41, 209–11) the same poet evokes the death of Adonis brought about by Ares in the form of a boar (for the Beroe episode, cf. the analyses of Gennaro D'Ippolito, *Studi nonniani. L'epillio nelle Dionysiache* [Palermo, 1964], pp. 110–14). The version Nonnos qualifies as recent could well be a Phoenician version whose conflation with the myth of Poseidon and Amymone is not its least interesting feature. Though Paleolithic hunters have nothing to do with this, another Adonis is nonetheless taking shape. He does so in a geographical domain where, more so than in the garden of Attica, Greek exegesis should take account of the Eastern narratives of which Aphrodite's lover is the incontestable heir (cf. n. 160 on Adonis and ripe fruits).
27. *Schol. T. in Il.* 24, 31.
28. *Metamorphoses* 10, 520–739.
29. Ibid., 537–41.
30. Etruscan mirror from West Berlin (Staatliche Museen, Antikenabteilung, Inv. Fr. 146): J. D. Beazley, "The World of Etruscan Mirrors," *Journal of Hellenic Studies* 69 (1949): 12–13 (a good photograph of it can be found in W. Attalah, *Adonis dans la littérature et l'art grecs* [Paris, 1966], p. 65, fig. 5).
31. Cf. R. Schilling, *La Religion romaine de Vénus* (Paris, 1954), pp. 165–67.
32. As opposed to the interpretation defended by Beazley, "The World of Etruscan Mirrors," n. 32.
33. Cf. Detienne, *Les Jardins d'Adonis*, p. 130, n. 1 [English trans., p. 67, n. 45]. In his essay on Panyassis of Halikarnassos (*Text and Commentary* [Leiden, 1974], pp. 120–25), Victor J. Matthews ascribes this trait to the version developed by Panyassis [fr. 25 k (b)].
34. Cf. Marcel Detienne, "L'olivier, un mythe politico-religieux," *Revue de l'Histoire des Religions*, no. 3 (1970), pp. 18–19.

35. *Metamorphoses* 10, 564–66.

36. So in *Schol. Theocr.* 3, 40 d; *Schol. Eur. Phoen.* 150; and modern mythographers have ratified the split (cf. W. Immerwahr, *De Atalanta* [Berlin, 1885]). This analysis, along with a series of extensions, was elaborated during seminars at the Ecole des Hautes Etudes en Science sociales in 1974–75. At the moment when its writing was largely completed, Giampiera Arrigoni of the University of Milan communicated to me the results of his research on the ambiguities of virginal status in the mythical tradition. If our analyses cross at several points, they diverge in their orientation. Mine is focused on the Ovidian confrontation between Adonis and Atalanta, while Arrigoni has chosen to view in Atalanta's mirror the ideological variations that develop around the story of a woman liberated from marriage and in doubt about masculine values.

37. *Metamorphoses* 10, 317–430.

38. Ibid., 322–23.

39. Lilly G. Kahil, "Autour de l'Artémis attique," *Antike Kunst* 7 (1964): 20–33.

40. Theocritus, *Epithalamion for Helen.*

41. Pausanias, 5, 16, 2.

42. Ibid. 3, 9, 2.

43. Aelianus, *VH* 13, 1; *Anecdota Graeca*, ed. I. Bekker, 1:444, 30–445, 13.

44. Apollodorus 3, 9, 2.

45. Theognis, 1289–94. The elegy's first verse (1283) invokes the injustice in love of an *erómenos* who flees his lover but cannot avoid being wounded or "penetrated." That there is a justice (*díkē*) in amorous relations has been shown by Gentili (cf. *Studi classici e orientali* (Pisa) 21 (1972): 60–72; and M. G. Bonnanno, "Osservazioni sul tema della 'giusta' reciprocità amorosa da Saffo ai comici," *Quaderni Urbinati di Cultura Classica* 16 (1973): 11–20.

46. In the space of six verses, the word *télos* appears explicitly three times and implicitly a fourth time through its equivalent *hōraîos.*

47. Cf. Detienne, *Les Jardins d'Adonis*, pp. 218–19 [English trans., pp. 117–18]. There is a *télos gámoio* (*Odyssey* 20, 74) or a *gamélion télos* (Aeschylus, *Eumenides* 835).

48. Theognis, 1289: anainoménēn *gámon* andrôn/pheúgein~1294: *télos* d'égnō kaì mál' anainoménē.

49. Hesiod, *Works and Days* 72; 76.

50. Pausanias, 9, 17, 3. Cf. Ch. Picard, "Athéna Zōstéria," *Revue des Etudes anciennes* 34 (1932): 245–53. Warriors are called *Zōstêres Enyoûs*: Callimachus, *Hymn to Apollo* 85.

51. *Parthenië Zónē*: *Odyssey* 11, 245.

52. *Paroemiographi Graeci*, 2, 513, 5–8 Leutsch-Schneidewin. At Troizen, young girls offered their girdle to Athena *Apatouria* on the eve of their marriage: Pausanias, 2, 32, 2; S. Wide, *De Sacris Troizeniorum Hermionensium Epidauriorum* (Uppsala, 1888), pp. 15 ff. Cf. J. Boardman, "Ionian Bronze Belts," *Anatolia* 5 (1960): 179–89, and P. Schmitt, "Athéna Apatouria et la ceinture: les aspects féminins des Apatouries à Athènes," *Annales E.S.C.* 32 (1977): 1059–73.

53. *Zōstêr* of Ares: Apollodorus 2, 5, 9; after having obtained it from the queen of the Amazons, Herakles comes to the temple of Mycenae to offer it to Hera, sovereign goddess of marriage: Euripides, *Herakles* 416–18. On the definitions respectively of *zostêr* and *mitrē* (a girdle worn next to the skin), cf. C. Rolley, *Les Statuettes de bronze (Fouilles de Delphes*, vol. 5) (Paris, 1969), pp. 20–21.

54. Parmenides, 8, 4 and 8, 32; 42.

55. *Kreobóroi*: Aeschylus, *Suppliants* 287.

56. The epithet is given her by Nonnos, *Dionysiaca* 35, 82, but the *Iliad* bestows it on the Amazons (2, 189; 6, 186). The history of its double interpretation in antiquity from Aeschylus to Aristarkhos has just been written by T. Drew-Bear, "Imprecations from Kourion," *Bulletin of the American Society of Papyrologists* 9 (1972): 88–92.

57. Apollodorus 3, 9, 2.

58. Ibid.: *kathōplisménē*; Hyginus, *Fabulae* 185. For an epic model mentioning the stake on which the loser's head was fixed and showing Atalanta armed and running, see J. Schwartz, *Pseudo-Hesiodeia* (Leiden, 1960), pp. 362 ff.

59. A. Minto, "La corsa di Atalante e Hippomenes figurata in alcuni oggetti antichi," *Ausonia* 9 (1919): 78–86.

60. Pausanias, 3, 12, 4 ff.; 13, 6.

61. Pindar, *Pythian Odes* 9, 105–24.

62. *Metamorphoses* 10, 557–59.

63. *Schol. Lycophr.*, ed. Scheer, 266, 4–21.

64. Apollodorus 3, 14, 4; *Scholia to Euripides, Hippolytus* 1421.

65. *Schol. Lycophr.* 831; Nonnos, *Dionysiaca* 42, 209–11. Cf. Attalah, *Adonis,* pp. 320–21, where the husband exacting vengeance from the seducer is named Hephaistos.

66. Erika Simon, "Aphrodite und Adonis. Eine neuerworbene Pyxis in Würzburg," *Antike Kunst* 15 (1972): 20–26, pl. 5–7.

67. E. Langlotz, "Aphrodite in den Garten," *Sitz. Heidelberg. Akad. d. Wissenschaften, Phil. hist. Klasse,* 1953–54, 2, Heidelberg, 1954.

68. K. Schefold, *Untersuchungen zu den Kertscher Vasen* (Berlin, 1934), p. 103, figs. 41–42; Attalah, *Adonis,* pp. 203–4, fig. 60.

69. Simon, "Aphrodite and Adonis," p. 22.

70. Aeschylus, *Prometheus* 172.

71. Simon, "Aphrodite and Adonis," p. 21.

72. O. Keller, *Thiere des Classischen Alterthums in culturgeschichtlicher Beziehung* (Innsbruck, 1887), pp. 154–57.

73. Simon, "Aphrodite and Adonis," p. 22.

74. Ibid., p. 25.

75. A. Greifenhagen, *Griechische Eroten* (Berlin, 1957), pp. 40–46.

76. The pictorial documents reproduced by Simon, "Aphrodite and Adonis," pl. 6, clearly reveal this difference from the pyxis for which they are cited.

77. O. Keller, *Classischen Alterthums* 154–57; 389, n. 81 (other evidence).

78. Rich figurative documentation in H. Jereb, s.v. "Panther," *R.-E.* (1949), c. 767–76.

79. F. Wotke, ibid., c. 747–67.

80. Aelianus, *NA* 6, 2.

81. Its valor is asserted in one Homeric simile (*Iliad* 21, 573–80) and only one, but at the moment when Hektor's challenge is announced. All the Trojans are fleeing toward their city but Agenor faces him "like a panther coming from a deep thicket to face a hunter. His heart feels neither fear nor the urge to flee when he hears the bark of the hounds. If the man grazes or smites him first, even if transfixed by a spear, he does not forget his valor; he will first either attack or perish."

82. Aesop, *Fabulae* 42; Plutarch, *Moralia* 500 C–D.

83. Aelianus, *NA* 5, 54.

84. Theophrastus, *De Causis Plantarum* 6. 17. 9; Pliny, *HN* 8, 62; 21, 39.

85. [Aristotle], *Problemata* 13, 4, 907b 35–37.

86. Pliny, *HN* 13, 6. *Pardáleion* is the name of a *phármakon* also called "strangle-panther," *pardaliankhés*, which is a toxic plant of the genus *Aconitum* used to destroy the beast. But the animal's prudence must be taken into account. For as soon as the panther has recognized the poison smeared by men on the bait offered his greedy appetite, he sets out to find the only effective remedy— human excrement. The hunters' ruse is to place some in proximity to the trap, but in a suspended container, so that the panther exhausts himself trying to reach it without ever succeeding (Aristotle, *Hist. An.* 9, 6, 612a 5–12; [Aristotle], *Mir.* 6 in *Paradoxographi Graeci* 226, Giannini).

87. Aristotle, *Hist. An.* 9, 6, 612a 12–15. Cf. Theophrastus, *De Causis Plantarum* 6, 5, 2; [Antigone], *Mir.* 31 in *Paradoxographi Graeci* 50–51, Giannini.

88. Pliny, *HN* 8, 62.

89. Aelianus, *NA* 5, 40.

90. *Physiologus* 1, 16, ed. Sbordone, p. 60, 5 ff.

91. Nicole de Margival, *Le Dit de la panthère d'amors*, ed. H. A. Todd (Paris, 1883). "Li vraie pantière . . . qarist de se douche alaine" ["The true panther . . . cures with his sweet breath"] the wounded or sick creatures that draw near: Richard de Fornival, *Li Bestiaire d'Amours*, ed. C. Segre (Milan, 1957), p. 45, 1. Cf. A. J. Festugière, "Le Bienheureux Suso et la panthère," *Revue de l'Histoire des Religions* 191 (1977): 81–84.

92. Philostratus, *Life of Apollonius of Tyana* 2, 1–2. The passage is reproduced in the *Bibliotheca* of Photius, 324a 25–b18.

93. Cf. H. G. Horn, *Mysteriensymbolik auf dem Kölner Dionysosmosaïk* (Bonn, 1972), pp. 109 ff.

94. Oppian, *Cynegetica* 4, 320–53; Timotheus of Gaza, 11, ed. M. Haupt.

95. Damascius, *Vita Isidori* 97.

96. Aristophanes, fr. 478 Kock; *Lysistrata* 1014–15.

97. Detienne, *Les Jardins d'Adonis*, pp. 121–22 [English trans., pp. 62–63].

98. Xenophon, *Memorabilia* 3, 1, 1, 5 f. The professors are divided about Xenophon's revelations: some cry blasphemy and speak of apocrypha, others seriously explain that it is not impossible that Socrates' relations with this sort of woman sometimes went beyond academic admiration. The debate fills chapter 11 of A. Delatte, *Le Troisième Livre des Souvenirs Socratiques de Xénophon* (Liège and Paris, 1933), pp. 148–61.

99. Hesychios, s.v. "aphrodisía ágra"; *Paroemiographi Graeci* 2, 150, 5 Leutsch-Schneidewin.

100. Aristotle, *Hist. An.* 9, 9, 614a 26–28.

101. On the Amasis cotylus (Louvre A479, see n. 21), in the parade of gifts and amorous couples, a kneeling man holds erect on his left hand a leaping panther whose erotic significance is guaranteed by the figure's context (see fig. 3). One should add here, while awaiting an iconographic investigation of animals of the "panther" type, an Etruscan mirror from Leningrad (Hermitage B505) that shows a standing Aphrodite embraced by Adonis. Beside them, a young satyr, flanked by two panthers; on the right, seated, a winged woman named Zipna holds an *alabastron* of perfume and a spatula designed to spread its contents over head and body: Beazley, "The World of Etruscan Mirrors," pp. 11–12, fig. 13. On a banded cup by the Centaur painter (CVA British Museum, 2 GB 2, 15, 1a and b), pointed out to me by A. Schnapp, a fawn hunt on one side corresponds to the *pursuit by a panther* of a young boy on the other. Doubtless a lipped cup found at Cyprus (*Bulletin de Correspondance hellénique*,

1962, p. 296, fig. 7) belongs to the same erotic context. It shows a panther opposite a hunter with member erect who carries the traditional staff (*lagōbólon*) for hunting hares.

One should likewise add, from among the documents analyzed by Henri Metzger, *Les Représentations dans la céramique attique du IVᵉ siècle* (Paris, 1951), pp. 47–53, a hydria from the Hermitage (cat. Ant. Bosph. 104h; Metzger no. 29, p. 49) on which a small panther appears under the seat of a young woman whose knees support a standing Eros. The erotic significance of the cat represented on several South Italian vases belongs to a symbolic system that goes from Eros to Dionysos. Here, the panther is just a more violent, more savage cat. Cf. T. M. C. Toynbee, *Animals in Roman Life and Art* (London, 1973), pp. 87–88. Claude Bérard has pointed out to me the cup by the painter of dogs and cats (Beazley, *ARV²* 866, 1; *Paralipomena* 426) reproduced by J. Dörig, *Art antique. Collections privées de Suisse romande* (Mayence, 1975), no. 215.

102. *Metamorphoses* 10, 698–704.

103. Ibid. 3, 9, 2.

104. P. Vidal-Naquet, "Le chasseur noir et l'origine de l'éphébie athénienne," *Annales E.S.C.* 23 (1968): 947–64; a version that is new in several points has been published in Italian in the anthology *Il mito. Guida storica e critica*, ed. Marcel Detienne (Rome and Bari, 1975), pp. 53–72, 245–52.

105. Klitias and Ergotimos, Volute crater 4209, Florence, Museo Archeologico, Side A, Scenes on the neck. J. Boardman, J. Dörig, W. Fuchs, and M. Hirmer, *L'Art grec* (Paris and Munich), 1966, pl. 92.

106. Aristophanes, *Lysistrata* 785–796.

107. Hesiod, fr. 76 Merkelbach-West. Cf. Schwartz, *Pseudo-Hesiodeia*, pp. 361–66.

108. Hesiod, fr. 76, 5–6 Merkelbach-West.

109. Ibid. 8–10.

110. Apollodorus 3, 9, 2.

111. *Metamorphoses* 10, 609–80.

112. Ibid., 637.

113. Ibid., 676–78.

114. Theocritus, 3, 40–42. A goatherd afflicted by his lady-love's coldness recounts, by way of amorous incantation, a series of adventures in which love miraculously triumphs. His third example is the mad passion (*lússa*) of Aphrodite for Adonis: though he is but a corpse, she cannot resolve herself to part from his bosom. Learned attention is focused on this bosom. Is it maternal or erotic? Unless one is seriously convinced, to begin with, of the great goddess's generosity in welcoming to her bosom her son and lover, one will hesitate to follow the "maternal" interpretation of C. Segal, "Adonis and Aphrodite," *L'Antiquité classique* 38 (1969): 82–88.

115. J. Trumpf, "Kydonische Apfel," *Hermes* 88 (1960): 14–22. D. R. Littlewood, "The Symbolism of the Apple in Greek and Roman Literature," *Harvard Studies in Classical Philology*, 1967, pp. 148–68; M. Lugauer, *Untersuchungen zur Symbolik des Apfels in der Antike* (Erlangen, 1967); I. Chirassi, *Elementi di culture precereali nei miti e riti greci* (Rome, 1968), pp. 73–90.

116. *Metamorphoses* 10, 644–48.

117. *Schol. Theocr.* 3, 38 b; Servius, *In Verg. Aen.* 3, 13; *Ecl.* 6, 61. A third version derives the fruit from the crown of Dionysos, from which they retain their power as erotic charms: Philetas fr. 18 Powell.

118. Athenaeus, 3, 84c.

119. Pausanias, 2, 17, 4.

120. Philostratus, *Life of Apollonius of Tyana*: 4, 28.

121. Pherekydes *apud FGrH* 3 fr. 16 Jacoby. Cf. Euripides, *Hippolytus* 742–50: the bridal chamber of Zeus and Hera stands upon the fertile earth that bears the golden trees of the Hesperides.

122. *Poetae Melici Graeci*, 286 Page. "Pomegranates for the newlyweds" is a proverb used of the most beautiful gifts destined for the most beautiful beings: *Paroemiographi Graeci* 2, 770 Leutsch-Schneidewin.

123. P. Zancani-Montuoro, "Note sui soggetti e sulla tecnica delle tabelle di Locri," *Atti e Memorie della Società Magna Grecia* 1 (1954): 71–106 (particularly pp. 98–99).

124. *Poetae Melici Graeci*, 187 Page.

125. *Solon, Nomoi: die Fragmente der Gesetzwerkes*, ed. E. Ruschenbusch (Weisbaden, 1966), frs. 127a, b, c.

126. Plutarch, *Conjugal Precepts* 138 D.

127. Women celebrating the Thesmophoria were forbidden to eat grain that had fallen on the ground: Clement of Alexandria 2, 19, 3 Stählin; 15, 9. Cf. L. Deubner, *Attische Feste* (Berlin, 1932; reprint. ed. 1956), p. 58.

128. *Homeric Hymn to Demeter* 372–74; 411–13 in N. J. Richardson's edition (Oxford, 1974), with the notes on pp. 276–77 and 286–87.

129. H. Gaidoz, "La réquisition d'amour et le symbolisme de la pomme," *Annuaire de l'Ecole Pratique des Hautes Etudes, Sciences historiques et philologiques* (Paris, 1902), pp. 5–33.

130. *Schol. Aristoph. Nubes* 996–97.

131. Theocritus, 5, 88.

132. Antoninus Liberalis, *Métamorphoses* 1, ed. M. Papathomopoulos. Likewise, the loves of Akontios and Kydippe: Callimachus, frs. 67–75 Pfeiffer.

133. *Homeric Hymn to Aphrodite* 69–74.

134. Plutarch, *Quaest. Rom.* 264B; Diodorus, 5, 73, 2–3; Pausanias 2, 32, 7–9; Pollux, 3, 38; *Etymologicum Magnum*, s.v. "*gamēlía.*"

135. Ovid, *Metamorphoses* 10, 681–82.

136. Ibid., 698–707. Cf. *Mythographi Vaticani* 1, 3ʋ; 2, 47.

137. Apollodorus 3, 9, 2; Hyginus, *Fabulae* 185, 6. In Nonnos's version, *Dionysiaca* 12, 87–89, the goad, *oîstros* (at once "gad-fly" and "mad passion"), is sent down by Artemis, as Gianpiera Arrigoni has pointed out to me. This time the "punishment" of Atalanta comes from the divine power who cultivates virginity and the hunt. But Artemis's game is subtle: to confine Atalanta to the forests and to chastity, she has recourse to Aphrodite's weapon, the sex drive.

138. *Metamorphoses* 10, 694.

139. Herodotus, 2, 64.

140. *Metamorphoses* 10, 698–707. Cf. 547–52.

141. Hyginus, *Fabulae* 185; *Mythographi Vaticani* 1, 39; Servius, *In Verg. Aen.* 3, 113.

142. We refer here to the research of Gianpiera Arrigoni (see n. 36).

143. Pliny, *HN* 8, 43. Cf. Apollodorus, *Library*, ed. J. G. Frazer (1921), 1: 401, 2.

144. In interpreting the antierotic character of lions, perhaps one should adduce the castration of Attis provoked by the anger of Cybele, as G. Devallet of the Université Paul-Valéry has suggested to me.

145. Herodotus, 3, 108–9.

146. Ibid., 108.

147. Aristotle, *Hist. An.* 6, 31, 579 b1–5.

148. The hare is an animal with a great deal of sperm. The proof of it, says

Aristotle, is its abundant hair: "it is the only animal that has hair beneath its feet and inside its jaws" (*Gen. An.* 4, 5, 774a 32–35).

149. Philostratus, *Imagines* 1, 6, 5.

150. Stamnos from London, E440 (c. 480 B.C.): Greifenhagen, *Grieschische Eroten*, fig. 25.

151. Cf. Attalah, *Adonis*, p. 202, fig. 58.

152. Aristotle, *Gen. An.* 3, 1, 750a 30–35. Cf. III, 10, 760b 23.

153. *Metamorphoses* 10, 725–39.

154. Ibid., 728–31. Aphrodite addresses Persephone: "Once you were able to change a woman's body into fragrant mint, *in olentes vertere mentas;* but I myself would be blamed for offering the son of Cinyras a new form." The version to which Ovid alludes is that in which Mintha, dismembered by Persephone, is at Hades' request transformed into garden mint (*Schol. vetera in Nic. Alex.* 375, ed. Keil; Strabo, 8, 3, 14. Cf. Detienne, *Les Jardins d'Adonis* p. 143 [English trans., p. 72]).

155. *Metamorphoses* 4, 190–255. The whole story of Leucothoe and Clytia should be integrated into the analysis of Ovid's narrative about Myrrha and about Adonis: Sun, seduced by Leucothoe, disguises himself as *mother* to have union with the daughter of the king of the Land of Spices; Clytia, Sun's abandoned wife, makes public the scandal, thus provoking Leucothoe's death and condemning Clytia herself to pine away until she is transformed into heliotrope, an odorless flower always turned toward the sun. Paolo Fabbri, who set forth this analysis during a seminar at the Ecole des Hautes Etudes, drew attention to the fact that the disconsolate Clytia froze herself into "Thesmophoric" behavior: seated on the bare earth . . . for nine days without water or food . . . until her limbs [were] attached to the ground (*Metamorphoses* 4, 256–70).

156. Ibid., 732; 4, 250. "The flower never wilts but at a breath of wind, a fact that also won it its name," Pliny, *HN* 21, 165.

157. *Metamorphoses* 10, 732–39.

158. To pronounce words that are like the wind: *anemōlia bázein* (*Odyssey* 11, 465). And Lucian once allowed himself the expression "the anemones of speech," *anemōnai tōn lógōn,* words borne on the wind (Lucian, *Lexiphanes* 23). For some the name of the anemone is linguistically a derivative of *ánemos:* P. Chantraine, *Dictionnaire étymologique de la langue grecque* (Paris, 1968), 1: 86. In his lexicon (Latte's edition, 4882) Hesychios records a broad definition of anemone: "any plant quickly destroyed by the winds."

159. J. André, *Lexique des termes de botanique en latin* (Paris, 1956), s.v. "anemōnē"; K. Lembach, *Die Pflanzen bei Theokrit* (Heidelberg, 1970), pp. 167–68.

160. Dioscorides, 2. 176. 3. In the analysis by Giulia Piccaluga, which is centered on the misfortunes of the hunter, the anemone plays an important role. It confirms decisively that even in death Adonis is denied access to the kingdom of Demeter. For proof Piccaluga has recourse to the testimony of Ammianus Marcellinus (22, 9, 15) which, by her lights, puts the death of Adonis in relation to the appearance of a variety of anemone called *adonis,* whose flowering season precedes the wheat harvest (*ormai prossima maturazione del grano*). Here is the text she appeals to: "Evenerat autem eisdem diebus annuo cursu completo, Adonea ritu veteri celebrari amato Veneris, ut fabulae fingunt, apri dente ferali deleto *quod in adulto flore sectarum est indicium frugum.*" The Emperor Julian arrives at Antioch. "On those days, the year having gone its course, they were celebrating the festival of Adonis in the ancient manner; Venus's lover, as the myths say, had perished beneath the savage blows of a

boar, which means that he is the symbol of wheat harvested in the flower of youth" (in his edition *Le Storie di Ammiano Marcellino* [Turin, 1965], A. Selem refers to 1a, 1, 11 for the expression *in adulto flore*). There is not a trace of the anemone, but instead an interpretation that transforms Adonis into a symbol of ripe fruits, in the manner of Porphyry (*Perì agalmátōn*, 7, p. 10* in *Vie de Porphyre*, ed. J. Bidez [Ghent, 1913]).

In a careful review (*Revue de l'Histoire des Religions* 155 [April 1974]: 208–11), R. Turcan observes that Adonis goes from *áhōros* to *hōraîos* and that in dying prematurely he also dies in the full blossom of youth, in the flower of beauty. It is not only a possible interpretation, but Porphyry and others have already made it. Adonis's relationship to Demeter, as well as the balance of all the concepts inventoried in different configurations, is thereby modified. I am totally in agreement with R. Turcan (211) when he says that my analyses principally bear on information and versions of the myth in an Athenian context between the sixth and fourth centuries B.C.; likewise when he recognizes the necessity for other readings that would specify the evident variations during the Hellenistic period and furthermore distinguish the geographical allegiance of these other versions (Alexandria or Byblos), as H. Seyrig suggests ("Antiquités syriennes, 96. La résurrection d'Adonis et le texte de Lucien," *Syria* 49 [1972]: 100).

161. Theophrastus, *Historia Plantarum* 6, 8, 1–2; Pliny, 21, 64.

162. *Schol. Lycophr.* 831, ed. Scheer 266, 22f.

163. [Bion], *Lament for Adonis* 65–66.

164. Theocritus, 5. 92; *Paroemiographi Graeci* 3, 635, 1–2 Leutsch-Schneidewin.

165. Plutarch, *Sympos.* 3, 1.3, 648A.

166. Theophrastus, *De Causis Plantarum* 6 .5. 1: roses and artificial perfumes (cf. Detienne, *Les Jardins d'Adonis*, p. 50 [English trans., p. 24]).

167. *Schol. Theocr.* 5. 92e Wendel.

168. I append a word or two to answer the objection addressed me in advance by the very ones who revealed on my behalf the interest of the anemone metamorphosis (J. P. Borle in a review appearing in *Museum Helveticum* 30 [1973]: 236–37; and A. Voelke, by correspondance on this subject). On several occasions, botanists and medical writers (Dioscorides, 2, 176, 2; Pliny, 21, 165; Galen, 11, 831) relate that among its other virtues this plant causes lactation and brings on menstruation. The effectiveness of the anemone in menstruation and lactation would orient Adonis's flower toward feminine fertility and should, as a consequence, attach it to the world of Demeter. First of all it is worth noting that in the most detailed version, which is Ovid's, the explicitly mentioned characteristics of the plant deny any relationship to fertility. If the text of the *Metamorphoses* gave no detail concerning the anemone, it would doubtless be necessary to gather up the maximum amount of information dispersed in the culture in order to envisage the different solutions that could account for the choice of the flower in this mythical context. However, one and the same plant can be the vehicle for highly contrasting values and its *position* in a myth or group of myths does not necessarily permit differentiation between such values.

In the mythical ensemble centered on perfumes and seduction the case of the anemone is not isolated. Another investigation enabled me to complement the lettuce of Adonis with the lettuce of Hera; I was able to show that although this vegetable caused impotence, it was likewise responsible for the birth of Hebe, or Youth, the daughter whom Hera engenders on her own by eating lettuce. The same is true of the *agnus castus*, or chaste tree, in the format of the Thesmophoria. Its anaphrodisiac characteristics are of a piece with its effectiveness in the phenomena of menstruation and lactation. For these two botanical species, the apparent contradiction in their double connotations depends on a deeper split between the plane of erotic desire, sexual pleasure, and seduction, on the one hand, and the plane of feminine fertility and the production of legitimate children, on the other. If the split is the same in both cases, it does not apply to the same object. For lettuce, it is the couple: lettuce makes the man impotent, depriving him of desire, while for the woman, it is the substitute for sperm and seed. As for the chaste tree, its effects concern the woman alone: the plant that clamps shut her body's desire is also the one that, by bringing on the menses and causing milk to mount, favors reproductive activity in her.

In this perspective, the double orientation of the anemone's values could be explained by invoking the two sexes: it would connote the halt of seduction (without perfume and without fruit) and the death of masculine erotic desire while leaving intact and ready a kind of female fertility held in ignorance of any form of sexuality. The myth that exploits the anemone's second register of meaning remains to be found.

169. Detienne, Les Jardins d'Adonis, pp. 192–94 [English trans., pp. 102–4].

170. Cf. n. 158.

171. Plato, Symposium 196b: Love never settles on that which does not flower or has lost its flower, be it body, soul, or whatever.

172. Nicander, frs. 65 and 120 Schneider.

173. Cf. the schematic classification in Detienne, Les Jardins d'Adonis, pp. 33–36 [English trans., pp. 13–15].

174. At the birth of the divine child in the Fourth Eclogue, as around Dionysos, fragrant flowers mingle with the rarest perfumes. The Dionysiac Golden Age can only be realized by running their gamut from one extreme to the other. Cf. H. Jeanmaire, Le Messianisme de Virgile (Paris, 1930), pp. 180 ff.; J. Hubaux and M. Leroy, "Vulgo nascetur amomum," Mélanges J. Bidez (Ann. Inst. Philol. Hist. Or. 2 [1933–34]) (Brussels, 1934), pp. 505–30.

175. Plutarch, Sympos. 3. 1. 646B.

176. The split marked out by Baudelaire in the tercets of Correspondances is alien to Greece, which ignored perfumes of animal origin such as musk and likewise did not associate sensuality with the mysticism of perfumes of vegetal origin, such as incense and benzoin. Cf. J. Gengoux, "Les Tercets des Correspondances de Baudelaire: signification et prolongements," Annales de la Faculté des Lettres et Sciences Humaines. Université de Dakar, 1971, pp. 119–87. Empedocles may have a trope in which corruption is the reverse of a perfume's violence. The richest emanations arise from the flower and from maturity, but the more powerful the scent, the more life is consumed, the more death threatens (J. Bollack, Empédocle [Paris, 1965], 1: 235–37). The documentation gathered by Saara Lilja, The Treatment of Odours in the Poetry of Antiquity (Commentationes Humanarum Litterarum 49) (Helsinki, 1972), should lead to further exploration of the Greek's olfactory categories.

177. Theognis, 1275–77.

178. Cf. Detienne, Les Jardins d'Adonis, p. 225, n.3 [English trans., pp. 122 and 177, n. 113 (reading supra p. 102, not 122)]. In Aphrodite's tale Ovid suggests the same ambiguity in speaking of Atalanta: "Hippomenes sees Atalanta's face and her body stripped of its veils, a body such as mine or yours would be if you became a woman, si femina fias" (Metamorphoses 10, 579). My thanks to G. Devallet for bringing this passage to my attention.

Chapter 3

The *Nouvelle Revue de Psychanalyse*, which devoted its sixth number to the "Destinies of Cannibalism," published a first version of this essay (1972, pp. 231–46).

1. A. Lang, *Myth, Ritual, and Religion* (New York, 1887).

2. Much material is gathered and analyzed from various viewpoints by Marie Delcourt, "Tydée et Mélanippe," *Studi e Materiali di Storia delle Religioni* 37 (1966): 139–88; Giulia Piccaluga, *Lykaon. Un tema mitico* (Rome, 1968); W. Burkert, *Homo Necans* (Berlin and New York, 1972).

3. Hesiod, *Theogony* 459–60.

4. Cf. Jean-Pierre Vernant, "Mètis et les mythes de souveraineté," *Revue de l'Histoire des Religions* 3 (1971): 29–76 (particularly pp. 41–44), reprinted in Marcel Detienne and Jean-Pierre Vernant, *Les Ruses de l'intelligence. La mètis des Grecs* (Paris, 1974), pp. 70–74.

5. G. Mihailov, "La légende de Térée," *Annuaire de l'Université de Sofia* 50, no. 2 (1955): 77–208.

6. As in Welcker, Hiller von Gaertringen, Cazzaniga . . .

7. These myths have their place in an interpretation of the mythology of honey.

8. Cf. Marcel Detienne, "Orphée au miel," in *Faire de l'histoire*, ed. Jacques Le Goff and Pierre Nora (Paris, 1974), 3: 56–75.

9. Cf. Marcel Detienne, *Les Jardins d'Adonis: La mythologie des aromates en Grèce* (Paris, 1972), pp. 71–113 [*The Gardens of Adonis: Spices in Greek Mythology* trans. Janet Lloyd (Sussex, 1977), pp. 37–59], and the introduction to it by Jean-Pierre Vernant, pp. vii–ix, xl–xliii [English trans. pp. v–viii, xxix–xxxii], as well as the analyses of Pierre Vidal-Naquet, "Chasse et sacrifice dans l'Orestie d'Eschyle," *Parola del Passato*, 1969, pp. 401–25 (reprinted in Jean-Pierre Vernant and Pierre Vidal-Naquet, *Mythe et tragédie en Grèce ancienne* [Paris, 1972], pp. 135–58; "Valeurs religieuses et mythiques de la terre et du sacrifice dans l'*Odyssée*," *Annales E.S.C.* 25 (1970): 1278–97; idem, "Bêtes, hommes, et dieux chez les Grecs," in *Hommes et bêtes. Entretiens sur le racisme*, ed. L. Poliakov (Paris, 1975), pp. 129–42.

10. Porphyry, *De Abstinentia* 1, 6; Aristotle, *Politics* A, 8, 1256b 7–26. Cf. P. Moraux, *A la Recherche de l'Aristote perdu. Le dialogue "Sur la Justice"* (Paris and Louvain, 1957), pp. 100–107; M. Laffranque, *Poseidonios d'Apamée* (Paris, 1964), pp. 468 and 478, along with the remarks of A.-J. Voelke, *Studia Philosophica* 26 (1966): 287.

11. Porphyry, *De Abstinentia* 1, 4.

12. Epicurus, *Rar. Sent.* 32 in Usener, *Epicurea* 78, 10–14, and the analyses of P. Moraux (n. 10). Cf. likewise J. Méleze-Modrzejewski, "Hommes libres et bêtes dans les droits antiques," in Poliakov, ed., *Hommes et bêtes*, pp. 75–102.

13. Hesiod, *Works and Days* 276–78; Plato, *Politicus* 271d; *Protagoras* 321a. But the orthodoxy of Hesiod and Aristotle is answered by the fable on the animal world in which the eagle's *húbris* infringes on equity, on the *díkē* the fox expected from him when he goes into partnership with the bird of the upper air (Archilochos, 168–74 Lasserre and Bonnard). For the Christian tradition, cf. J. Passmore, "The Treatment of Animals," *Journal of the History of Ideas* 36 (1975): 195–218.

14. Porphyry, *De Abstinentia* 1, 13.

15. Detienne, *Les Jardins d'Adonis*, p. 33 [English trans. p. 13].

16. Vernant, introduction to Detienne, *Les Jardins d'Adonis*, pp. xxxvi–vii [English trans., pp. xxvii–viii].

17. T. Cole, *Democritus and the Sources of Greek Anthropology* (Princeton, 1967), pp. 6, 20–21, 150.

18. Cf. A. J. Festugière, "A propos des arétalogies d'Isis," *Harvard Theological Review* 42 (1949): 216–20, reprinted in *Etudes de religion grecque et hellénistique* (Paris, 1972), pp. 145–49.

19. Thucydides, 3, 94.

20. Ephoros in *FGrH* 70 fr 42.

21. Herodotus, 4, 106.

22. Aristotle, *Nichomachean Ethics* 7, 1148b 19–25; *Politics* 8, 1338B 19–22.

23. Plato, *Laws* 782b6–c2.

24. Anonymous commentary in *Comment. in Aristot. graeca* (Berlin, 1892), 20: 427, 38–40.

25. Jean-Pierre Vernant, "Ambiguïté et renversement. Sur la structure énigmatique d'*Oedipe-Roi*," in Vernant and Vidal-Naquet, *Mythe et tragédie en Grèce ancienne*, pp. 116–17; 128–30.

26. Plato, *Republic* 571c. Cf. 619b–c.

27. Herodotus, 3, 25. Cf. Jean-Pierre Vernant, "Les Troupeaux du Soleil et la Table du Soleil," *Revue des Etudes grecques* 85, no. 2 (1972): xiv–xvii.

28. Porphyry, *De Abstinentia* 2, 57.

29. J.-P. Waltzing, "Le crime rituel reproché aux chrétiens du IIe siècle," *Bulletin de l' Académie royale de Belgique, Cl. Lettres*, 1925, pp. 205–39.

30. Cf. Detienne, *Les Jardins d'Adonis*, pp. 78–111 [English trans., pp. 40–59].

31. Cf. ibid., p. 99, n. 2 [English trans., p. 149, n. 98].

32. This interpretation is developed below, pp. 68–94.

33. J. Schmidt, s.v. "Omophagia," *R.-E.* (1939), c. 380–82; H. Jeanmaire, *Dionysos* (Paris, 1951), p. 256.

34. Euripides, *Bacchae* 1185–89.

35. Ibid., 1240–42.

36. Plutarch, *Quaest. Graec.* 38, 299E; Antoninus Liberalis, *Metamorphoses* 10. For all the existing material, see I. Kambitsis, *Minyades kai Proitides* (Jannina, 1975).

37. Aelianus, 3, 42.

38. Euripides, *Bacchae* 1674.

39. Porphyry, *De Abstinentia* 2, 8.

40. J. Haussleiter, *Der Vegetarismus in der Antike* (*RGVV*, vol. 24) (Berlin, 1935), pp. 167–84.

41. The comparison between Cynics and Hippies seems pertinent at certain points: E. Shmueli, "Modern Hippies and Ancient Cynics: A Comparison of Philosophical and Political Developments and Its Lessons," *Cahiers d'Histoire Mondiale* 12 (1970): 490–514.

42. "Make life savage" (*tòn bíon apothēriōsai*) as Plutarch puts it, *De Esu Carnium* 995 C-D, speaking of Diogenes the Cynic.

43. Diogenes Laertius, 6, 56; 105; Julian, *Disc.* 7, 214C; Dio Chrysostom, 6, 62, ed. G. de Budé, 1, 120, 16–21; 6, 21–22, ed. G. de Budé, 1, 111, 4–11.

44. Dio Chrysostom, 6, 25, ed. G. de Budé 1, 111, 23–28; Plutarch, *Aqua an ignis util.* 2, 956B. Cf. Cole, *Democritus*, pp. 150–51.

45. Dio and Plutarch, ibid.

46. Dio Chrysostom 10, 29–30, ed. G. de Budé, 1, 145, 12–16.

47. Theophilus, *Ad Autolycum* 3, 5 (Von Arnim, *SVF*, 3, fr. 750).

48. K. von Fritz, s.v. "Pythagoreer," *R.-E.* (1963), c. 214–19.

49. G. Méautis, *Recherches sur le Pythagorisme* (Neuchâtel, 1922), pp. 10–18; and W. Burkert, *Weisheit und Wissenschaft* (Nuremberg, 1962), p. 194 [rev. and trans. Edwin L. Minar, Jr., *Lore and Science in Ancient Pythagoreanism* (Cambridge, 1972), 198 ff.].

50. Aristophon, *Pythagoristes*, fr. 3 in *Fragmenta comicorum graecorum*, 3, 362, ed. Meineke; Alexis, *Pythagorizusa*, frgs. 2 and 3 in *FCG*, 3, 474–75; Aristophon, ibid., fr. 1 in *FCG*, 3, 360–61.

51. Antiphanes, *Mnemata* in *FCG*, 3, 87.

52. The *pēra* (Antiphanes, ibid., uses the synonym *kórukos*) and *tríbōn* (Aristophon, ibid., frs. 4 and 5 in *FGC*, 3, 362–63).

53. Cf. P. Tannery, "Sur Diodore d'Aspende," in his *Mémoires scientifiques* (1925), 7: 201–10. Although Burkert, *Weisheit und Wissenschaft*, pp. 192–99 [English trans., pp. 198–205] has well observed the relationship between these Pythagoreans in Comedy and a personage like Diodoros of Aspendos, he has paid more attention to continuity since the *acusmatici* than to the rupture attested to by this type of Pythagorean-Cynic, this marginal figure whom the city, this time, rejects and who is radically divorced from the group and society once constituted by the Pythagoreans.

54. Stratonikos in Athenaeus, 163 E–F.

55. Sosikrates in Athenaeus, ibid.

56. Alexis, *Tarentini*, frgs. 1 and 2 in *FCG*, 3, 483.

Chapter 4

This analysis, which was presented and discussed in several seminars at the Scuola Normale Superiore of Pisa, has been published in the *Annali della Scuola Normale Superiore di Pisa, Cl. di Lettere e Filosofia*, S.III, vol. 4, no. 4 (1974): 1193–1234. The present version has been corrected at many points.

Among those who did me the kindness of discussing these pages, I wish to thank, in particular, Luc Brisson, Walter Burkert, Jean-Louis Durand, Pierre Smith, and Froma Zeitlin.

1. C. Bérard, *Anodoi. Essai sur l'imagerie des passages chthoniens (Bibl. helv. rom.)* 1974, pp. 103–11.

2. For the older school of thought, see A. Loisy, *Les Mystères païens et le mystère chrétien*, 2d ed. (Paris, 1930). Today's tone is set by A. D. Nock, "Hellenic Mysteries and Christian Sacraments" (1952), reprinted in *Essays on Religion and the Ancient World*, ed. Z. Stewart (Oxford, 1972), 2: 791–820. The same author's volume *Early Gentile Christianity and Its Hellenistic Background* (New York, 1964) has appeared in a mediocre French translation under the title *Christianisme et hellénisme* (Paris, 1973).

3. V. Macchioro, *Zagreus. Studi intorno all' orfismo* (Florence, 1930).

4. A. J. Festugière, "Les mystères de Dionysos (2)," *Rev. biblique*, 1935, p. 381 [*Etudes de religion grecque et hellénistique* (Paris, 1972), p. 47].

5. Dionysos is present throughout the book René Girard has devoted to sacrifice: *La Violence et le sacré* (Paris, 1972) [*Violence and the Sacred*, trans. Patrick Gregory (Baltimore, 1977)].

6. Festugière, *Etudes de religion grecque et hellénistique*, p. 76.

7. We intend to show that this information is in accord with the basic discourse of Greek mysticism as two previous studies have tried to define it: "La cuisine de Pythagore," *Archives de Sociologie des Religions* 29 (1970): 141–61

(cf. *Les Jardins d'Adonis: La mythologie des aromates en Grece* (Paris, 1972), pp. 71–114 [*The Gardens of Adonis: Spices in Greek Mythology*, trans. Janet Lloyd (Sussex, 1977), pp. 37–59]; and "Entre bêtes et dieux," *Nouvelle Revue de Psychanalyse* 6 (1972): 231–46 (see chapter 3 in this volume). Against the hyper-critical attitude of Wilamowitz, Linforth, and Moulinier, the traditional point of view has been defended by Guthrie, Lagrange, and several others, such as H. J. Rose. Two studies by K. Prumm give an account of the arguments of both camps: "Die Orphik im Spiegel der neueren Forschung," *Zeitschrift für katholische Theologie*, 1956, pp. 1–4, and s.v. "Mystères. L'Orphisme" in *Dictionnaire de la Bible*, suppl. 6, 1957, c. 55–86.

8. The evidence is presented by Festugière, "Les mystères de Dionysos (2)," pp. 376–79, and especially by I. M. Linforth, *The Arts of Orpheus* (Berkeley and Los Angeles, 1941), pp. 307–64. Cf. *Orphicorum Fragmenta*, ed. O. Kern, 2d ed. (Berlin, 1963), fr. 34–36, 209–14.

9. [Aristotle], *Problems*, 3, 43 ed. Bussemaker, 4, 331, 15 ff.; S. Reinach, "Une allusion à Zagreus dans un problème d'Aristote," in *Cultes, Mythes et Religions* (Paris, 1923), 5: 61–71.

10. *Orph. Frag.* 35 Kern.

11. Aristophanes, *Frogs* 1032; Plato, *Laws* 782c [*Orph. Frag., Test.* 212 Kern]; Euripides, *Hippolytus* 952–53 [*Orph. Frag., Test.* 213 Kern].

12. Cf. D. Sabbatucci, *Saggio sul misticismo greco* (Rome, 1965), pp. 69–83.

13. Cf. Detienne, "La cuisine de Pythagore," pp. 141–62.

14. We have borrowed them from Sabbatucci, *Saggio sul misticismo greco*, pp. 87–126.

15. Plato, *Laws* 782c.

16. Pierre Boyancé, "Platon et les cathartes orphiques," *Revue des Etudes grecques* 55 (1942): 217 ff. The bibliography can be found in his *Le Culte des Muses chez les philosophes grecs* (Paris, 1936; reprint ed. 1972), p. 375.

17. Herodotus, 2, 81; cf. E. des Places, *La Religion grecque* (Paris, 1969), p. 199.

18. Aristophanes, *Frogs* 1032: *Orpheùs mèn gàr teletàs th' hēmîn katédeixe phónōn t' apékhesthai.*

19. Cf. J. Casabona, *Recherches sur le vocabulaire des sacrifices en grec, des origines à la fin de l'époque classique* (Aix-en-Provence, 1966), pp. 160–62, and the information given in Detienne, *Les Jardins d'Adonis*, pp. 79–80 [English trans., p. 41].

20. Particularly Linforth, (*The Arts of Orpheus*, pp. 68–72), who gives three possible meanings of *phónos* in this context: murder, cannibalism, and meat. But, as we shall see later, the last of these meanings subsumes the other two. *Contra*, F. Graf, *Eleusis und die orphische Dichtung Athens in vorhellenistischer Zeit* (Berlin and New York, 1974), pp. 31–39.

21. This interpretation is already to be found in Firmicus Maternus, *De errore profanarum religionum* 6, p. 15, 2 Ziegler [*Orph. Frag.* 215 Kern].

22. H. Jeanmaire presented their arguments and gave the decisive critique of the thesis (*Dionysos: Histoire du culte de Bacchus* [Paris, 1951], pp. 371–78). In his book *The Greeks and the Irrational* (Berkeley and Los Angeles, 1951, p. 155) E. R. Dodds persists in believing that this myth about Dionysos "is founded on the ancient Dionysiac ritual of *Sparagmos* and *Omophagia*."

23. W. F. Otto (*Dionysos*, 2d ed. [Frankfort, 1933], p. 120) did not mistake the difference between the myth and omophagy.

24. A lynching scene is what Girard imagines, (*La Violence et le sacré*, pp. 190 and 347). He is thus able to justify his key interpretation, the "scapegoat"

interpretation. The same confusion between two ways of eating has led several historians of Orphism to recognize the Titans devouring Dionysos on a hydria from the British Museum (E246): a bearded Dionysos wearing a leafy crown stands by while a child is torn to pieces by a clothed figure in a Thracian locale. Cf. C. Smith, "Orphic Myths on Attic Vases," *Journal of Hellenic Studies* 11 (1890): 343–51; A. B. Cook, *Zeus* (1914) 1: 654, pl. 36; W. K. C. Guthrie, *Orpheus and Greek Religion*, 2d ed. (New York, 1966), p. 130. Only Henri Metzger, *Les Représentations dans la céramique attique du IV* siècle (Paris, 1951), p. 263, n. 3, has realized with Sir John Beazley's help the improbability of such an interpretation. At the same time, H. Jeanmaire wished to consider the gesture "Dionysos" makes in reaction to the scene a sign of disapproval aimed directly at certain orgiastic excesses; the Orphic critique is then only an aspect of this disapproval (*Dionysos*, p. 407).

25. J. Rudhardt, *Notions fondamentales de la pensée religieuse et actes constitutifs du culte dans la Grèce classique* (Geneva, 1958), pp. 259–61.

26. Aristophanes, *Peace* 960 and *Scholia*; Plutarch, *De def. or.* 435C; 437A–B; idem, *Quaest. conv.* 729F; *Scholia* to Apollonius Rhodius, 1. 425; Porphyry, *De Abstinentia* 2, 9.

27. These several objects, "cone, top, dice, mirror," are the ones listed in the Gourob Papyrus, a document of the third century B.C. that describes a ritual of initiation into the mysteries of Dionysos. The ritual consists in the sacrifice and eating of a goat and a ram, the drinking of a beverage by the initiate, and the manipulation of these "toys," symbols of the passion of Dionysos: *Orph. Frag.* 31 Kern. Cf. M. J. Lagrange, *Les Mystères: l'Orphisme* (Paris, 1937), pp. 113–17; Jeanmaire, *Dionysos*, pp. 472–73.

28. In Nonnos's version (*Dionysiaca* 6, 172–73 [*Orph. Frag.* 209 Kern]), Dionysos receives the fatal blow at the moment he looks upon his own face, apparently distorted by the surface of the mirror. So Dionysos is caught in the trap of his own image; for the Neoplatonic philosophers, Dionysos is cut to pieces because he is attached to his own physical appearance, because he is caught in the snare of his material being. In the *Poimandres* of the *Corpus hermeticum* (1, 14), the primordial man is bent over the earth; he has seen a form like his own reflected in the water, has begun to love it, and has desired to live near it. Cf. J. Pépin, "Plotin et le miroir de Dionysos," *Revue internationale de Philosophie*, 1970, pp. 304–20.

29. *Orph. Frag.* 209 Kern. Herodotus, 2, 41, shows clearly that the *mákhaira* is, along with the cauldron and the spit, the tool that signifies Greek eating behavior to a foreigner. The invention of the *mákhaira*, in fact, signals the death of the plow ox (Aratos, *Phainomena* 131 and Plutarch, *De Esu Carnium* 2, 998A).

30. Pausanias, 7, 37, 8.

31. *Orph. Frag.* 34–36 and 210–11 Kern.

32. Cf. in particular *Orph. Frag.* 35 Kern. The choice of the verb for tearing to pieces doubtless has deeper implications: the critique of a specifically Dionysiac procedure, as Jeannie Carlier suggests to me (cf. below, 91–93).

33. Plato, *Euthydemus* 301a.

34. A. J. Festugière, who neglects these details, as do other interpreters of Orphism, wishes to see them as the chattering one-upmanship of a late tradition (*Etudes de religion grecque et hellénistique*, p. 44). The Aristotelian *Problem*, of which he seems unaware, proves on the contrary the antiquity and importance of these apparently useless details. Declaring that the double cooking of the meat is a confusion, A. Henrichs (*Die Phoinika des Lollianos* [Bonn, 1972], p. 68) easily persuades himself that the sacrifice of the child Dionysos

by the Titans is a faithful reflection of the "Dionysiac ritual" he believes he has discovered in the romanesque tale of Lollianos. It describes a group of confederates sharing the heart of a child during a nocturnal masquerade. Without trying to discuss an interpretation that disregards all reference to Orphism, that neglects any distance between Dionysiac religion and Orphism, and that maintains that drinking wine is a way of giving life to the child "Dionysos," whose heart has just been eaten in a communal, initiatory meal, it should at least be noted that in Lollianos's tale the heart is the only part of the victim that is consumed, while in the myth of Dionysos eaten by the Titans, the only part of the child that does not disappear into his assassins' stomachs is, precisely, the heart (cf. below, n. 89).

35. The expression *tà legómena en têi teletêi* led S. Reinach ("Une allusion à Zagreus," pp. 64–65) astray into the Eleusinian Mysteries. But Pierre Boyancé, "Remarques sur le salut selon l'orphisme," *Revue des Etudes anciennes* 43, (1941): 160–61, is correct in considering it a reference to a work of Orpheus that is sometimes called *Teleté* (*Orph. Frag.* 34 Kern: Orpheus, poet of the *Teleté*), sometimes *Teletaí* (*Orph. Frag.* 301 and *Test.* 223 Kern). Aristophanes' fundamental testimony (*Frogs* 1032) closely associates the *teletaí* Orpheus taught us and the refusal of the carnivorous life for which he is responsible.

36. The version in Clement of Alexandria (*Orph. Frag.* 35 Kern) is in perfect agreement with the Aristotelian *Problem: Kathêpsoun próteron, épeita obelískois*.

37. This is what the sacrificers in Menander's *Dúskolos* do when they have forgotten to take along the indispensable cauldron (ll.456 ff. and 519).

38. One can cite, among others, the sacrificial meals performed by the Cyclops, Lykaon, and Atreus. In *Cyclops* 243–47, 356 ff., Euripides presents his ogre first tasting pieces of flesh "hot off the charcoal," then devouring the rest once it had been "softened by cooking in the cauldron." This monstrous meal's antisacrificial character is all the more clearly marked by its blatant use of the fundamental features of the model sacrifice. When Lykaon offers his guest a human victim in order to prove the divine nature of his visitor, he proceeds as an expert sacrificer: "He softened in boiling water one share of the victim's palpitating limbs, the other he roasted on the fire (Ovid, *Metamorphoses* 1, 228–29). The same technique is used by Atreus: "He uses spit as well as pot" (Seneca, *Thyestes* 1060–65). The reference to sacrifice is further emphasized by the name of the beneficiary of this feast of human flesh: *Thyestes* (*Thuestés*) seems to signify "the sacrifice-man" (K. Kérenyi, *Heroen der Griechen* [Zurich, 1958], p. 327).

39. F. Sokolowski, *Lois sacrées d'Asie Mineure* (Paris, 1955), no. 50, 1.34.

40. G. Ricci, "Una Hydria ionica da Caere," *Annuario della Scuola archeologica di Atene* 24–25 (1946–48): 47–57, pl. 4 (1–2–3).

41. Aristotle, *De partibus animalium* 667b 1 ff. and 673b 15 ff. The internal organs are themselves subdivided into *splánkhna* and *éntera*, the latter being the entrails, i.e., the organs of the lower abdominal cavity (Plato, *Timaeus* 73a). My thanks to Guy Berthiaume for these details.

42. Aristotle, *De partibus animalium* 674a 4–6.

43. Ibid. 673b 1–3.

44. Ovid, *Fasti* 3, 805 ff.

45. Plutarch, *Camillus* 5, 5–6. Cf. J. Hubaux, *Rome et Véies* (Paris, 1958), pp. 221–85.

46. Athenion *apud* Athenaeus, 14, 660E [*Com. Graec. Frag.* 3, 369 Kock].

47. Theophrastos *apud Schol. AD in Il.* 1, 449.

48. Cf. A. Delatte, "Le Cycéon, breuvage rituel des mystères d'Eleusis," *Bulletin de l'Acad. Roy. de Belgique, Cl. Lettres Sc. Mor. et Polit.* 40, 5ᵉ serie (1954): 691–93.

49. F. Sokolowski, *Lois sacrées des cités grecques* (Paris, 1969), no. 18, col. A, ll. 40–43.

50. G. Daux, "La Grande Démarchie: un nouveau calendrier d'Attique (Erchia)," *Bulletin de Correspondance hellénique* 87 (1963): 629.

51. *Odyssey* 3, 5–66.

52. Aristophanes, *Peace* 1115.

53. Athenaeus, 9, 410A–B.

54. *IG*, 9, I,¹ 330 [L. Lerat, *Les Locriens de l'Ouest* (Paris, 1952), 2: 252].

55. Aristotle, *Meteorologica* 4, 3, 380a36 ff.; 381a27 ff.; *Problems* 5, 34, 884a36 ff.

56. Philokhoros, *FGrH* 328 F173 Jacoby. Cf. Detienne, *Les Jardins d'Adonis*, pp. 210–11 [English trans., pp. 104–5].

57. Plato, *Republic* 372d–373a.

58. Ibid., 404 a–d.

59. Cf. above, p. 74.

60. Iamblichus, *De Vita Pythag.* 154, p. 87, 6–7 Deubner.

61. Detienne, "La cuisine de Pythagore," pp. 155–57.

62. Another form of inversion would have been to boil what should be roasted and to roast what should be boiled. But all the victim's parts are treated in the same way without any distinction between *splánkhna* and *kréa*. The victim is first divided into portions or slices (*moîrai* or *mérides*) as though all were to be boiled. But once removed from the cauldron, everything is roasted over the fire; normally only the *splánkhna* are so roasted. There is not, then, a simple inversion with respect to the ordinary sacrifice, but a double one. As Pierre Smith has pointed out to me, in one case the culinary processes are distributed among different parts of the victim; in the other, they are added cumulatively to the same parts of the victim.

63. *Orph. Frag.* 209 Kern.

64. Dio Chrysostom, 30, 55.

65. Jeanmaire, *Dionysos*, p. 390. The way had already been shown by J. Harrison, *Prolegomena to the Study of Greek Religion*, 3d ed. (Cambridge, 1922; reprint ed. New York, 1955), pp. 491–94.

66. This objection was made by M. P. Nilsson in a review of Jeanmaire's book: *Gnomon* 25 (1953): 276.

67. The Titans are nonetheless not the enemies *par excellence* of Dionysos, as Pierre Boyancé suggests ("Le *Dionysos* de M. Jeanmaire," *Revue philosophique*, 1956, p. 116) in trying to reduce this part of the myth to a dominant theme in the mythology of Dionysos: denial and persecution.

68. R. Martin, *Manuel d'architecture grecque* (Paris, 1965), 1: 425.

69. A. Orlandos, *Les Matériaux de construction et le technique architecturale des anciens grecs* (French translation by Vanna Hadjimichali) (Paris, 1966), 1: 136–48.

70. The relation between the Titans and plaster (*títanos*) has already been demonstrated, if briefly, by A. Dieterich, *Rheinisches Museum* 48 (1893): 280; Harrison, *Prolegomena*, pp. 491f., and in *Themis*, 15 and 17; Loisy, *Les Mystères païens et le mystère chrétien*, p. 33, n. 2.

71. M. Pohlenz, "Kronos und die Titanen," *Neue Jahrbücher für das klassiche Altertum* 37 (1916): 581–83, drew attention to these traditions.

72. *FGrH* 328 fr. 74 Jacoby (with his commentary in 2, b (*Supplements*) (Leiden, 1954), 1: 354–55).

73. *FGrH* 334 fr. 1 Jacoby (with his remarks, ibid. 627).

74. Pausanias, 2, 11, 5.

75. Aristotle, *Meteorologica* 4, 11, 389a28.

76. *Iliad* 2, 735.

77. Eustathius, 332, 24 ff.

78. The Bronze men are born from ash trees, and the Korybantes spring from the earth and grow like trees before the wondering eye of the Sun; or again they are the Dryopes, companions of the oak. Cf. Marcel Detienne, "L'Olivier, un mythe politico-religieux," *Revue de l'Histoire des Religions* 151 (1970): 15.

79. Hesiod, *Works and Days* 61; Xenophanes, B 33 Diels-Kranz, 7th ed.; Aristophanes, *Birds* 686; Pausanias, 10, 4, 3.

80. Plato, *Protagoras* 320D.

81. Pausanias, 7, 37, 5 [*Test.* 194 Kern].

82. Before "chewing them up," according to *Orph. Frag.* 107 Kern (*tàs sárkas masōntai*).

83. *Contra*, J. Rudhardt, "Les mythes grecs relatifs à l'instauration du sacrifice: les rôles corrélatifs de Prométhée et de son fils Deucalion," *Museum Helveticum* 27 (1970): 1–15. The Prometheus myth in Hesiod was the object of penetrating analyses in the seminars of Jean-Pierre Vernant, who provides an outline of them in *Mythe et société en Grèce ancienne* (Paris, 1974), pp. 177–94.

84. This is the viewpoint of Wilamowitz and Festugière.

85. Resuming I. M. Linforth's critique, L. Moulinier summarized the various objections (*Orphée et l'Orphisme à l'époque classique* [Paris, 1955], p. 59).

86. Plutarch, *De Esu Carnium* 996 C [*Orph. Frag.* 210 Kern].

87. Dio Chrysostom, 30, 10. Cf. for birth from soot, *aithálē, Orph. Frag.* 220 Kern. On a philological level, it has long since been customary to answer the objection by adducing the testimony of Xenocrates (fr. 20 Heinze), Plato's pupil, on the "Titanic" nature of the prison (*phrourá*) to which men are confined. Cf. Boyancé, "Remarques sur le salut selon l'Orphisme," pp. 167–68, and W. Burkert, *Homo Necans* (Berlin and New York, 1972), p. 249, n. 43.

88. *Orph. Frag.* 35 and 36 Kern.

89. *Orph. Frag.* 210 a and b Kern.

90. Cf. in general J. Keil, *Kulte im Prytaneion von Ephesos, Anatolian Studies Buckler* (Manchester, 1939), pp. 119–28.

91. Cult rules from the third century A.D., published by F. Miltner and G. Marech, *Anzeiger A. W. Wien* 96 (1959): 39–40; see also F. Sokolowski, *Lois sacrées des cités grecques. Supplément* (Paris, 1962), no. 121.

92. The meaning of *kardiourgoúmena* is clear from Hesychius, s.v. *kardioústhai*, where *kardiourgeîn* is glossed by *kardioulkeîn* "draw out the victim's heart," a meaning given by Lucian, *De Sacrificiis* 13. Cf. Henrichs, *Die Phoinika des Lollianos*, pp. 71–72.

93. There is no reason to prefer one to the other.

94. Published by S. Accame, *Mem. Ist. Stor. Arch. di Rodi* 3 (1938): 69–71, and available in Sokolowski, *Lois sacrés des cités grecques, Suppl.* no. 108, 1.1–3. J. and L. Robert, *Bull. Epig.*, 1946–47, no. 157, considered Pythagorean influence, but the rest of the inscription dissuaded them. It discusses paying specific sums into a trunk, a *thēsaurós*, to defray the cost of blood sacrifices with oxen, quadrupeds (pigs, goats, sheep), and roosters as victims. Though the presence of animal victims does not immediately exclude reference to Pythagoreans, the

mention of an ox as one of them seems to justify indeed the reservations voiced by J. and L. Robert.

95. Going barefoot and wearing white garments goes hand in hand with purity of hands and soul: P. Roussel and M. Launey, *Inscriptions de Délos* (Paris, 1937), no. 2529, ll.16–17. Whether this is a rule for life or just a temporary ascesis is discussed by H. Jeanmaire, "Sexualité et mysticisme dans les anciennes sociétés hellénistiques," in *Mystique et continence* (Les Etudes carmélitaines) (Paris, 1952), p. 54.

96. Cf. Pausanias, 1, 37, 4 [*Test.* 219 Kern/and *Orph. Frag.* 291 Kern].

97. Detienne, *Les Jardins d'Adonis*, pp. 96–100 [English trans., pp. 49–52].

98. Iamblichus, *Protrepticus* 21, 108, 5–6 Pistelli; Aristotle, fr. 194 Rose (Aelianus, *VH* 9, 17 and Aulus Gelluns, 4, 11); Iamblichus, *De Vita Pythag.* 109, in Deubner's edition 63, 2–3.

99. Plutarch, *Quaest. conv.* 2, 3, 1, 635 e–f. In the *Theolog. arithm.* of Pseudo-Iamblichus, c. 22, the brain is called the *arkhē* of man and the heart the *arkhē* of the living being.

100. Diogenes Laertius, 8, 28, in A. Delatte's edition, 127, 11–15, and see the comments of Delcourt, "Tydée et Mélanippe," *Studi e Materiali di Storia delle Religioni* 37 (1966): 176–78.

101. A. Olivieri, "L'uovo cosmogonico degli Orfici," *Memoria R. Acc. Arch. Lett. Napoli* 7, 1919 (1920): 297–334; R. Turcan, "L'Oeuf orphique et les quatre éléments," *Revue de l'Histoire des Religions* 159–60 (1961): 11–23.

102. Published by J. Keil in *Anzeiger Osterr. Akad. Wiss.*, 1953, pp. 16 ff., and available in Sokolowski, *Lois sacrées de l'Asie Mineure*, no. 84.

103. Letting the heart be consumed "on the sacred altars" means forbidding men to eat it and leaving it as part of the divine portion of the sacrifice.

104. Cf. G. Daux, "L'interdiction rituelle de la menthe," *Bulletin de Correspondance hellénique* 81 (1958): 1–5, who proposes, following Kalleris, to read *hēdeosmoú* instead of *ēdeosmoú*, i.e., to read one of the names of mint, a plant still used in Greece as an almost indispensable spice in the preparation of broad beans.

105. Ll.15–16 of the inscription (cf. n. 102).

106. A. D. Nock, in Stewart, ed., *Essays on Religion and the Ancient World*, p. 852.

107. The reasons given by Plutarch, *Quaest. conv.* 2, 3, 1, 635e–f, must be taken seriously. M. Tierney, "A Pythagorean Tabu," *Mélanges E. Boisacq* (Brussels, 1938), who tries to explain the Pythagorean tabu by the influence of Orphism, misunderstands the difference between the two mystic movements.

108. Cf. R. S. Harris, *The Heart and the Vascular System in Ancient Greek Medicine from Alcmaeon to Galen* (Oxford, 1973).

109. Philo, *Legum allegoriae* 2, 6, Mondésert 107. The value of this text was pointed out by Pierre Boyancé, "L'Apollon solaire," *Mélanges J. Carcopino* (Paris, 1966), p. 167, n. 3.

110. Aristotle, *De Partibus animalium* 670 a 23–26. Cf. S. Byl, "Note sur la place du coeur et la valorisation de la *mesótēs* dans la biologie d'Aristote," *L'Antiquité classique*, 1968, pp. 467–76.

111. Ibid., 666a 7–10 and 20–22.

112. Ibid., 666a 14–16, 666b 6–10.

113. Philolaos B 17. Cf. W. Burkert, *Weisheit und Wissenschaft* (Nuremberg, 1962), pp. 248 ff. [*Lore and Science in Ancient Pythagoreanism*, trans. E. L. Milnar (Cambridge, 1972), pp. 268 ff].

114. Aristotle, *De Gen. An.* 734a16 [*Orph. Frag.* 26 Kern]. Binding and weav-

ing are basic operators in the myths of the birth of man and the world. Cf. Marcel Detienne and Jean-Pierre Vernant, *Les Ruses de l'intelligence. La Mètis des Grecs* (Paris, 1974), p. 132 and n. 14.

115. Lagrange, *Les Mystères: l'Orphisme*, pp. 127–32; Guthrie, *Orpheus and Greek Religion*, pp. 100 ff.; Detienne and Vernant, *Les Ruses de l'intelligence*, pp. 128–31.

116. *Orph. Frag.* 61, 85, 107, 170, 237 Kern.

117. *Orph. Frag.* 207 Kern.

118. The "reformation" point of view was defended by E. Rohde, *Psyche*, 8th ed., trans. W. B. Hillis (London and New York, 1925; reprint ed. New York, 1966), 2: 335 ff., and criticized especially by Jeanmaire, *Dionysos*, pp. 396 ff. The thesis that the myth transforms omophagy into a criminal scenario was developed by M. P. Nilsson, "Early Orphism and Kindred Religious Movements," *Harvard Theological Review* 28 (1935): 203–4 [*Opuscula Selecta*, 2, 1952, 654–55].

119. Cf. above, pp. 61–64, as well as Vernant's introduction to Detienne's *Les Jardins d'Adonis*, pp. xli–xlii [English trans., pp. xxix–xxxii].

120. Porphyry, *De Abstinentia* 2, 8.

121. Plutarch, *Quaest. Graec.* 38, 299E–300A. There is pursuit, *díōxis*, and the weapon is a sword, *xíphos*, not a sacrificial knife.

122. J. Schmidt, s.v. "Omophagia," *R.-E.* (1939), c. 380–82.

123. Sokolowski, *Lois sacrées d'Asie Mineure*, no. 48, 1.2; the formula is explained by Festugière, *Classica et Mediaevalia*, 1956, pp. 31–34 [*Etudes de religion grecque et hellénistique*, pp. 110–13].

124. Herodotus, 3, 110. Cf. 2, 146, and 2, 97, which are cited as parallels by J. Hubaux and M. Leroy, "*Vulgo nascetur amomum*," *Mélanges J. Bidez* (*Ann. Inst. Philol. Hist. Or.* 2 [1933–34]) (Brussels, 1934), pp. 505–30.

125. Dionysius Periegeta, 5, 935–47, G. Bernhardy, ed., *Geographi graeci minores* (Leipzig, 1878), 1: 51 ff. Hubaux and Leroy, "*Vulgo nascetur anomum*," relate this scene to the birth of a Divine Child in the *Fourth Eclogue* of Vergil.

126. Euripides, *Bacchae* 702–68. The "ambiguities of the Golden Age" were brought out in connection with this text and several others by P. Vidal-Naquet, "Le mythe platonicien du 'Politique' et les ambiguïtés de l'âge d'or et de l'histoire," in *Langue, Discours et Société. Pour E. Benveniste* (Paris, 1975), pp. 374–90.

127. This interpretation of the iron and bronze was advanced by J. Roux, "Pillage en Béotie," *Revue des Etudes grecques* 76 (1963): 37. Cf. Euripide, *Les Bacchantes*, ed. and trans. with commentary by J. Roux (Paris, 1972), 2: 482–83.

128. Antoninus Liberalis, *Metamorphoses* 10, ed. M. Papathomopoulos; Plutarch, *Quaest. Graec.* 38, 299 E–F; Aelianus, *VH* 3, 42.

129. Cf. Jeanmaire, *Dionysos*, pp. 399–414.

130. Ibid., p. 401.

131. In the investigation of the Dionysiac mysteries throughout Egypt that was instituted by an edict of Ptolemy IV Philopator, the presiding royal functionary is obliged to question all those who performed initiations into the mysteries. His purpose is to learn who transmitted the rites during the three previous generations and also to obtain information under seal about the holy teachings. Cf. P. Roussel, "Un édit de Ptolémée Philopator relatif au culte de Dionysos," *CRAI*, 1919, pp. 237–43, and for clarification of some details, G. Zuntz, "Once More the So-called 'Edict of Philopator on the Dionysiac Mysteries,'" *Hermes* 91 (1963): 228–39. The diversity of sacred teachings corresponds to the plurality of practices.

132. Plato, *Republic* 364e. Cf. Euripides, *Hippolytus* 954.

133. Aeschylus, fr. 83 Mette (cf. his commentary, 2: 138–39).

134. In an essay on the solar Apollo, Pierre Boyancé, "L'Apollon solaire," pp. 149–70, showed that the equivalence of Apollo and Helios was the result of learned speculations for which the Pythagoreans were partly responsible.

135. "Le grand prêtre de Thrace au long surpellis blanc," as Joachim du Bellay calls him.

136. Cf. F. Cumont, "La grande inscription bacchique du Metropolitan Museum," *American Journal of Archaeology* 37 (1933): 249.

137. Porphyry, *De Abstinentia* 2, 8.

138. As has often been noticed, Mainadism is a female activity. In the Hellenistic period men doubtless took part in the mysteries as initiates and with important functions. But even then women retained the presidency of the thiasos and played the role of mystagogue (cf. Festugière, *Etudes de religion grecque et hellénistique*, p. 19, n. 4). Yet in the college at Torre-Nova, which was of Greek origin, the members were predominantly males and the dignitary who led the cortège bore the title *Hérōs*. The first female personage, who walks beside him, is the *Dāidoûkhos*, the torch-bearer. There may have been feminine Pythagoreans, but it seems there were no Orphic women. The *Orpheotelestaí* of Plato are in fact exclusively masculine. When the Orphic texts break their scornful silence, it is only to repeat the formula: "Nothing more bitchy than a woman" (*Orph. Frag.* 234 Kern). Nor is it simply a vague misogyny: apparently the hatred of Hippolytus for women, marriage, and sex is the characteristic that best justifies his being taken for a tool of Orpheus by Theseus and the Athenian spectators. But what about Eurydice and her privileged relationship to Orpheus? In the absence of a more rigorous interpretation, see Marcel Detienne, "Orphée au miel," *Quaderni Urbinati di Cultura Classica* 12 (1971): 7–23 [*Faire de l'histoire*, ed. Jacques Le Goff and Pierre Nora (Paris, 1974), 3: 56–75].

139. To see therein—as does Wilamowitz, *Der Glaube der Hellenen*, 2d ed. (Basel, 1959), 2: 190—a proof that ancient Orphism was more Apollonian than Dionysian is to misuse historical appearances. M. P. Nilsson is wiser to read the episode in terms of conflicts and tensions between the two mystic movements (*Geschichte der griechischen Religion*, 2d ed. [Berlin, 1955], pp. 686–87). Cf. in the same vein Jeanmaire, *Dionysos*, p. 407. For the thesis of A. Kruger (*Quaestiones Orphicae* [Halis Saxonum, 1934], pp. 30–33), who wishes to discern two Orphisms, see the objections of Lagrange, *Les Mystères: l'Orphisme*, pp. 44–46.

140. *Orph. Frag.* 35 Kern.

141. *Orph. Frag.* 35 and 210 Kern (particularly *Schol. Lycophr.* 208, p. 98, 5 Scheer). Dionysos at Delphi: Jeanmaire, *Dionysos*, pp. 187–98, etc.

General Index*

Accame, S., 114 n.94
Adoniai, 51, 104 n.160
Adonis, 8, 22, 24, 26, 27, 30, 34, 35,
   36, 37, 40, 45–51, 96 n.14, 98 n.26,
   99 n.36, 102 n.114, 104 n.160
Aegyptos, 13
Aetolia, 58
Agaue, 62, 63
Agrionia, 89
áhōros, 105 n.160
aithálē, 114 n.87
akropólis, 87
Alcibiades, 39
Alexander, 36
allēlophagía, 55
Allelophagy, 57, 58, 60–63, 75, 83, 86,
   89
Amazons, 33, 99 n.53
Amenophis III, 13
Amphitryon, 32
Amymone, 14
Anchises, 36
André, J., 104 n.159
androphágoi, 58
anemónē, 50, 104 n.158
Anemone, 50, 51, 104 n.156, n.158,
   n.160, 105 n.168
ánemos, 50
Animals: domestic and wild, 56; vs.
   man, 56, 57, 62, 70
ánodmon, 50
Antaios, 33
Anteros, 36
Anthestērión, 76
Anthropogony (Orphic), 71, 80, 81–84
Anthropophagy. *See* Cannibals
   (cannibalism)
antiáneira, 33, 100 n.56
Antisystem, 56, 59, 60, 61, 64, 65
apátē, 41
aphrodisía ágra, 40
Aphrodite, 25, 26, 27, 30, 32, 34–37, 40–
   51, 102 n.114
Apollo, 44, 77, 92
[Apollodorus], 82
Apollonian, 92
Apple, 41–44, 46, 73, 102 n.117. *See
   also* Pomegranate
Apple of Cydon, 43. *See also* Apple;
   Pomegranate
Arcadia, 73

Ares, 33, 35, 99 n.53
Argos, 13, 33, 42
Aristarkhos, 26
arkhé, 86, 87
Arkhytas, 66
Armenia, 39
Arrigoni, G., 99n. 36, 103 n.137, n.142
Arsacus, 39
Artemis, 25, 26, 27, 30, 31, 32, 35, 44,
   46, 103 n.137
Ascra, 16
Ashes, 69, 80, 82
Ash trees, 114, n.78
Asklepios, 85
Atalanta, 25, 27, 30, 31, 32, 33, 34, 35,
   40, 41, 42, 44, 45, 46, 47, 49, 51, 52,
   99 n.36, 100 n.58, 103 n.137, 106
   n.178
atélesta télei, 32–33
Athena, 26, 32, 76, 85
Atlas, 42
Atropos, 30
Attalah, W., 98 n.30, 100 n.65, 104
   n.151
Atthidographers, 80
Attica, 81
Atreus, 112 n. 38
Autochthony, 81
Autopsy, 20

Bacchae, 90
Bacchants, 62, 63, 72, 92
Bacchus, 69
Barley grains, 73, 76
Barthes, R., 97 n.11, n.17
Bassarai, 92
Bassaroi, 63, 89
Baudelaire, C., 106 n.176
Bear(s), 25, 26, 27, 31, 44, 97 n.20, 98
   n.25
Beazley, J. D., 98 n.30, n.32, 101 n.101,
   111 n.24
Bee-Woman (Women), 55, 57
du Bellay, J., 117 n.135
Bérard, C., 102 n.101, 109 n.1
Bérard, V., 4, 12
Bernhardy, G., 116 n.125
Beroe, 98 n.26
Berthiaume, G., 112 n.41
Bestiality, 57, 58, 59, 61, 62, 88, 90, 91,
   92. *See also* Savagery

*Both indexes were prepared by the translators.

Beyond, 90
bíos orphikós, 61, 71
Black One, 41, 47
Boar, 26, 27, 30, 31, 34, 35, 41, 46, 49, 51
Boardman, J., 99 n.52, 102 n.105
Boas, F., 19
Boeotia, 16
Boiled (boiling), 9, 61, 69, 73, 74, 77–79, 83, 84, 88, 93, 111 n.34, 113 n.62
Bollack, J., 106 n.176
Bonnanno, M. G., 99 n.45
Borle, J. P., 105 n.168
Bouphónia, 9, 84, 94
Boyancé, P., 96 n.3, 110 n.16, 112 n.35 113 n.67, 114 n.87, 115 n.109, 117 n.134
Brain, 85, 115 n.99
Brauron, 31
Brelich, A., 24, 97 n.19
Briareus, 75
Broad bean, 60, 61, 85, 86
Bull, 91
Burkert, W., 97 n.18, 107 n.2, 109 n.49, 114 n.87, 115 n.113
Byl, S., 115 n.110
Byzantium, 37

Caere, 74
Calame, C., 95
Calydon (Calydonian), 30, 31, 40
Cambyses, 59
Camilla, 75
Cannibals (cannibalism), 9, 17, 53–56, 58–65, 76, 83, 84, 86, 89, 91. *See also* Allelophagy; Endocannibalism
Carlier, J., 111 n.32
Carnivores (carnivorous), 33, 56, 61, 86, 89, 112 n.35
Casabona, J., 110 n.19
Cauldron, 9, 69, 74, 77, 79, 82, 89, 91, 93, 111 n.29, 112 n.37
Cazzaniga, 107 n.6
Cereal economy, 22, 23
de Certeau, M., 18, 96 n.3, n.10
Chantraine, P., 104 n.158
Chaos, 71
Charites, 44
Chaste tree, 105 n.168
Cheetah, 36, 37. *See also* Panther
Chios, 62
Chirassi, I., 102 n.115
Christianity, 68
Church Fathers, 69
Cinnamon, 90
City (city-state), 12, 17, 25, 56–60, 62, 64–68, 72, 84, 89, 90, 94, 109 n.53

Clytia, 104 n.155
"Cold" society, 3, 6
Cole, T., 108 n.17
Commensality (common repasts of gods and men), 57, 60, 65, 81
Cone, 111 n.27
Cook, A. B., 111 n.24
Cooked, 73, 79
Cooking. *See* Boiled (boiling); Roasted (roasting); Sacrifice
Courtesan, 39, 40, 47
Cultural history, 57. *See also* Sacrifice, and cultural history
Cumont, F., 117 n.136
Cybele, 45
Cyclops, 112 n.38
Cydon. *See* Apple of Cydon
Cynicism, 17, 56, 59, 62, 64, 65
Cynics, 17, 55, 64–66, 108 n.41
Cyrenaica, 36

Danaids, 7, 13–15
Danaoi, 13
Danaos, 13, 33
Daux, G., 113 n.50, 115 n.104
Deer (does), 25, 27, 34, 38, 97 n.21
Delatte, A., 101 n.98, 112 n.48, 115 n.100
Delcourt, M., 2, 107 n.2, 115 n.100
Delos, 85
Delphi, 92
Demeter, 14, 22, 43, 51, 76, 90–92, 104 n.160, 105 n.168
Demeter-Rhea, 84
Déspoina, 73
Deubner, L., 103 n.107
Devallet, G., 103 n.144, 106 n.178
diaspáô (diasparássō), 73, 111 n.32
diasparagmós, 9, 72, 73
Dice, 111 n.27
Dieterich, A., 113 n.70
díkē, 16, 57, 99 n.45, 107 n.13
Diodoros of Aspendos, 66, 109 n.53
Diogenes of Sinope, 64, 65
Dionysiac religion, 17, 54, 56, 59, 62–65, 73, 88, 89, 91, 92
Dionysos Bacchios, 90
Dionysos Bromios, 86
Dionysos *Ōmēstḗs*, 84
Dodds, E. R., 110 n.22
Dogs, 65, 66
Dolls, 69, 73
Donkeys, 65
dôra, 41
Dorig, J., 102, n.101, n.105
Drew-Bear, T., 100 n.56
Dryopes, 114 n.78
Dumézil, G., 4, 95 n.6, n.7

Eagle, 107 n.13
Earth, 42
Egg, 71, 86
eis aphrodísia deleázein, 44
ekmērizómena, 85
emánē, 42
Endocannibalism, 47, 59, 61, 62, 65, 66
éntera, 112 n.41
Ephesos, 85, 92
Ephoros, 20
Epoptía, 20, 21
erastḗs, 25, 34
Erchia, 76
erómenos, 25, 34
Eros, 36, 37, 42, 48, 51, 52, 71
es bathùn érōta, 42
es méson, 58
Ethiopia, 90
Euboia, 32
Euboulos, 26
Eupatrids, 77
Eurytanes, 58
Eustathius, 81
Excrement, 54

Fabbri, P., 104 n.155
Fate, 30
Faure, P., 13, 14, 96 n.17
Felis Acinonyx, 37
Felis Panthera, 37
Festugière, A. J., 68, 101 n.91, 108 n.18,
    109 n.4, n.6, 110 n.8, 111 n.34, 114
    n.84, 116 n.123, 117 n.138
Finnish School, 13
First-Born, 71, 88
Footraces (footracing), 30, 31; and
    hunting, 33; and marriage, 33, 34;
    and sexual identity, 31
de Fornival, R., 101 n.91
Fox, 25, 38, 97 n.21
François Vase, 41
Frazer, J. G., 47, 69, 72
Fuchs, W., 102 n.105

Gaidoz, H., 103 n.129
gámos, 32, 33
Gap, 71
Garden of the Hesperides, 42
genéseōs arkhḗ, 85
Gengoux, J., 106 n.176
Gentili, B., 99 n.45
géras, 77
Gernet, L., 15
Girard, R., 109 n.5, 110 n.24
Girdle, 32, 33, 99 n.50, n.51, n.52, n.53
Goats, 60, 70, 90, 111 n.27, 114 n.94

Golden Age, 57, 60, 65, 79, 90–92, 106
    n.174, 116 n.126
Golden fruit, 41, 49, 102 n.117, 103
    n.121. See also Apple; Pomegranate;
    Quince
Graf, F., 110 n.20
Granet, M., 15
Guardians, 78
gúpsos, 80
Guthrie, W.K.C., 110 n.7, 111 n.24,
    116 n.115
Gypsum, 9, 69, 80, 82, 93

Hades, 43, 49
hágnos, 85
Hamilcar Barca, 59
Hares, 25, 27, 34, 41, 46, 48, 49, 97
    n.21, 102 n.101, 103 n.148
Harris, A. S., 115 n.108
Harrison, J., 113 n.65, n.70
Haussleiter, J., 108 n.40
Heart, 70, 84–87, 112 n.34, 115 n.103
Hearth, 87
hēdúsmata, 76
Hekataios, 21
Helen, 31, 43
Hellanikos, 5
Hellenists (Hellenism), 1–3, 10, 11, 14,
    16–19, 53
Henrichs, A., 111 n.34, 114 n.92
Hera, 14, 25, 42–44, 46
Herakles, 26, 40, 99 n.53
Herder, 18
Hermes, 35, 37
Hermokhares, 44
Herodotus, 20, 21
Hesiod, 11, 15, 16, 41, 44, 54, 57, 71,
    72, 81, 82
Hesperides, 43
Hesperides, Garden of the, 42
Hestia, 87
Hímeros, 48
Hippies, 108 n.41
Hippolyta, 33
Hippolytos, 25, 35, 117 n.138
Hippomenes, 33, 41, 42
Hirmer, M., 102 n.105
History, 13, 20–22, 96 n.1; cumulative,
    3, 4; stationary, 3, 4. See also Myths,
    historical analysis of
Homer, 81
Honey, 54, 55, 90
Honeymoon, 54, 55, 90
Hoopoe, 55
hōraíē, 32
hōraîos, 32, 105 n.160
Horn, H. G., 101 n.93

"Hot" society, 3, 6
Hubaux, J., 106 n.174, 112 n.45, 116
    n.124, n.125
húbris, 16, 107 n.13
Humanism (humanist), 18, 19
Hunter-gatherers, 22, 23
Hunt (hunters, hunting), 23, 24, 73,
    101 n.86; and farming, 24; and
    footracing, 31, 33, 44; and marriage,
    30–34, 41, 45; and masculinity, 34,
    40; and sex/seduction, 25–27, 35,
    37, 39, 40, 44, 47, 49, 51, 52
hydrophóroi, 14
Hyginus, 6

Immerwahr, W., 99 n.36
Incense, 50, 106 n.176
Incest, 65
Initiation rites, 24, 25, 80
Intermediary. See Mediator (mediation)
D'Ippolito, G., 98 n.26
Iroquois, 53
Isara, 33
íunx, 38

Jeanmarie, H., 80, 97 n.22, 106 n.174,
    108 n.33, 110 n.22, 111 n.24, n.27,
    113 n.65, n.66, 115 n.95, 116 n.118,
    n.129, n.130, 117 n.139, n.141
Jereb, H., 100 n.78
Jews, 59
Justin, 59

Kahil, L. G., 99 n.39
Kalleris, 115 n.104
Kambitsis, J., 98 n.24, 108 n.36
kardiourgoúmena, 85, 114 n.92
Karousou, S., 97 n.21
kasalbás, 39
katapínein, 54
kathōplisménē, 31, 100 n.58
Keil, J., 114 n.90, 115 n.102
Kekrops, 81
Keller, O., 37, 100 n.77
Kérenyi, K., 112 n.38
Khaleion, 77
Kirk, G. S., 5, 11, 12, 13, 18, 95 n.5, n.9
Kite, 75
Kithairon, 90
Knucklebones, 69
Kouretes, 85
kréa, 113 n.62
Kronos, 53, 54, 57, 88
Kruger, A., 117 n.139
kruptós, 24
Ktesylla, 44

Laffranque, M., 107 n.10
Lagrange, M. J., 110 n.7, 111 n.27, 116
    n.115, 117 n.139
Lamia, 58, 61
Lang, A., 107 n.1
Langlotz, E., 100 n.67
La Penna, A., 97 n.14
Launey, M., 115 n.95
Leach, E., 3, 95 n.5
Lembach, K., 104 n.159
Lemnian, 7
Leopard, 37, 47, 91
Lerat, L., 113 n.54
Lerne, 14
Leroy, M., 106 n.174, 116 n.124, n.125
Lesbos, 62
Lettuce, 26, 50, 51, 105 n.168
Leucothoe, 50, 104 n.155
Lévêque, P., 22, 97 n.13, n.14, n.15
Lévi-Strauss, C., 1–3, 7–10, 14, 18, 95
    n.1, n.2, n.3, n.4, 96 n.16, n.18
Libyan, 33
Lilja, S., 106 n.176
Linforth, I. M., 110 n.7, n.8, n.20, 114
    n.85
Lion(s) (lioness), 27, 30, 31, 34, 40,
    44–49, 51, 52, 91, 103 n.144
Littlewood, D. R., 102 n.115
lógos, 2
Loisy, A., 69, 109 n.2, 113 n.70
Lokris, 77
Lokroi, 43
Loraux, N., 96 n.15
loutrophóroi, 14
Lugauer, M., 102, n.115
Lykaon, 53, 54, 112 n.38
Lykourgos, 92
Lynching, 73, 110 n.24
Lysistrata, 39

Macchioro, V., 68, 109 n.3
mágeiros, 75
Magna Graecia, 66
Mainads, 62, 90–92
mákhaira, 73, 111 n.29
Man: vs. beasts, 56, 57, 62, 70, 84, 88,
    90, 93; vs. gods, 56, 57, 70, 72, 82,
    84, 88, 90, 93
Mannhardt School, 72
Marathon, 80
Marech, G., 111 n.91
de Margival, N., 101 n.91
Marriage, 10, 30–34, 41, 45, 62, 99 n.36.
    See also Hunt (hunters, hunting)
Martin, R., 113 n.68
Matthews, V. J., 98 n.33

Meat (non-visceral), 74–77
Méautis, G., 109 n.49
Mediator (mediation), 3, 49, 55, 57, 64, 65, 71, 82
Megalopsychia, 98 n.26
Mekone, 82
Melanion, 41, 47
Meleagros, 27, 30, 31, 41, 47
Mélèze-Modrzejewski, J., 107 n.12
mēloboleîn, 44
mêlon, 43
Memory, 35
Menelaos, 43
mēría, 76
Meslin, M., 95 n.5
Methydrion, 73
mêtis, 54
Mêtis, 54
Metzger, H., 102 n.101, 111 n.24
Meuli, K., 97 n.18
Mihailov, G., 107 n.5
Miletus, 74, 90
Milk, 90, 91
Miltner, F., 114 n.91
Minoan, 80
Mint, 50, 86, 104 n.154, 115 n.104
Mintha, 49, 50, 104 n.154
Minto, A., 100 n.59
Minyads, 62, 63, 89, 91
Mirror, 69, 73, 111 n.27, 111 n.28
mítrē, 99 n.53
Molpoí, 74
Moraux, P., 107 n.10, n.12
Moulinier, L., 110 n.7, 114 n.85
Mounikhia, 31
Mount Kynthos, 85
Mount Pangaion, 92
Müller, K.-O., 12, 21
Müller, M., 7
Muses, 35
Mustês (epithet of Dionysos), 68
mûthos, 2, 48
Myrrh, 8, 40, 51
Myrrha, 40
Myrrhina, 39
Myths, historical analysis of, 3, 4, 12–16, 23
Myths, structural analysis of, 1, 2, 16, 55, 93, 94; and ethnographic context, 2, 8, 14, 15; and historical change, 15–18; historical critique of, 22; and meaning, 9, 10; and narrative form, 10–12; philological critique of, 97 n.14; and planes of signification, 7, 8, 15, 24, 55; principles of, 5–7; and interpretation of symbols, 105 n.168; and syntax, 9. See also Mediator (mediation)

Mytheme, 2, 10
Mythographi Vaticani, 6

Nauplia, 13
Nectar, 50, 91
Neîkos, 71
Nenci, G., 96 n.2
Neoplatonist, 69, 111 n.28
nēphálios mékhri splánkhnōn, 76
Nestor, 76
Nicander, 26, 50, 51
Nietzsche, F., 18
Night, 88
Nilsson, M. P., 113 n.66, 116 n.118, 117 n.139
Nock, A. D., 86, 109 n.2, 115 n.106
Nonnos, 73
Nora, P., 96 n.1
Nysa, 39, 90

Octopus, 64
Oedipus, 2, 3, 7, 65
oîstros, 103 n.137
Olivieri, A., 115 n.101
Olympia, 31
Olympian, 68, 75
ōmádios, 62
ōmēstḗs, 62, 88
ōmophagía, 9, 110 n.22
ōmophágion embállein, 90
Omophagy, 57, 58, 62–64, 72, 73, 88–90, 92, 93, 116 n.118
Onomakritos, 81, 83
Oppian, 39
optà kréa, 78
Orach, 66
Origen, 59
Orlandos, A., 113 n.69
Orpheotelestaí, 71
Orpheus, 9, 61, 70, 72, 79, 83, 87, 89, 90, 92
Orphism, 17, 56, 60, 70, 71, 80, 82–84, 87–93; vs. Dionysiac religion, 88–93
Osiris, 69
Otto, W. F., 110 n.23
oulokhútai, 76
Ouranos, 88

Painter of Meidias, 48
Painter of Würzburg Amymone, 35, 37, 40
Paleolithic Age, 23, 97 n.18, 98 n.26
Pamphylia, 39
Pandora, 32
Panther, 37–40, 44, 47, 49, 62, 100 n.81, 101 n.86, n.91, n.101

pánthēr, 37
párdalis, 37
Pardalium, 38
Paris, 36
Parricide, 65
parthénos, 31
Passmore, J., 107 n.13
Pausanias, 13, 81
Peisistratos, 69
Peithō, 36, 37, 40
Peleus, 41
Penelope, 33
Pentheus, 62, 63
pepaínein, 78
pépansis, 78
Pépin, J., 111 n.28
péra, 109 n.52
Perdiccas, 98 n.26
Perfumes, 8, 38, 39, 50, 51, 56, 90, 106
    n.174, n.176
Persephone, 43, 44, 49, 88
Persuasion, 36, 37, 44, 49
pétasos, 35
Phanes, 71, 86–88
Pherekydes, 5
Philo, 86
Philokhoros, 78, 80
Philolaos, 87
Philologists, 4, 5, 97 n.14
Philosophy, 1
Philostratus, 39, 48
Phlious, 66
phónos, phóni, 61, 72, 83, 110 n.20
phrónēsis, 38
Piccaluga, G., 21, 97 n.14, n.16, 104
    n.160, 107 n.2
Picard, C., 98 n.26, 99 n.50
Pigs, 60, 70, 114 n.94
Pindar, 11, 12
des Places, E., 110 n.17
Planes of signification. See Myths,
    structural analysis of
Plaster, 69, 113 n.70. See also títanos
Plato, 65
Plow ox, 56, 60, 82, 84, 93, 111 n.29
Pohlenz, M., 113 n.71
poikilía, 38
poikílos, 38
Poluphontế, 26
Polytekhnos, 54, 55
Pomegranate, 32–44, 50, 51, 103 n.122.
    See also Apple
Pontic region, 58
pórdalis, 39
Porphyry, 63, 92
Poseidon, 14, 76
Pot. See Cauldron
Pouillon, J., 96 n.20, n.22

Proitos, 25
Prometheus, 9, 32, 57, 60, 61, 64, 65,
    70–73, 79, 82, 93, 114 n.83
Prōtógonos, 71
Prumm, K., 110 n.7
Ptolemy IV Philopator, Edict of, 116
    n.131
Punica Granatum, 42
Purification ritual, 77
Pythagoras, 17, 61, 66, 79
Pythagorean Cynics, 66, 67, 109 n.53
Pythagoreans (Pythagoreanism), 9, 17,
    18, 55, 56, 59, 60–62, 66, 67, 70, 79,
    84–86, 92, 93
Pylians, 76
Pylos, 76

Quicklime, 9, 80, 81, 93
Quince, 43, 44. See also Apple;
    Pomegranate

Raw, 57, 62–64, 73, 79, 88, 90. See also
    Omophagy
Reinach, S., 110 n.9
Renard, 38
Rhea, 54, 92
Rhegium, 43
rheûma tês odôdês, 50
Rhodes, 85
rhódos, 50
Rhombus, 69, 73
Ricci, G., 112 n.40
Richardson, N. J., 103 n.128
Roasted (roasting), 9, 61, 69, 73–75,
    77–79, 83, 84, 88, 93, 111 n.37, 113
    n.62
Robert, L., 96 n.4, n.5, n.6, n.7, n.8, n.9,
    n.10, 97 n.11, 114 n.94
Roebuck, 62
Rohde, E., 116 n.118
Rolley, C., 99 n.53
Rooster, 65, 114 n.94
Rose, 50, 51
Rose, H. J., 110 n.7
Rotten, 54
Roussel, P., 115 n.95, 116 n.131
Roux, J., 116 n.127
Rudhardt, J., 111 n.25, 113 n.83
Running. See Footraces (footracing)

Sabbatucci, D., 17, 110 n.12, n.14
Sacrifice: alimentary blood, 9, 10, 17,
    56, 57, 61, 69, 70–79, 81–85, 88–91,
    94; and cultural history, 75, 76, 78,
    79, 83

Sale, W., 97 n.20
Salted/unsalted, 75–77, 79
Samaran, C., 96 n.4
Sarapis, 85
sárx, 75
Savagery, 53, 57, 64, 75, 91
Schauenburg, K., 97 n.21
Schefold, K., 100 n.68
Schepens, G., 96 n.2
Schilling, R., 98 n.31
Schmidt, J., 108 n.33, 116 n.122
Schmitt, P., 99 n.52
Schnapp, A., 101 n.101
Schwabl, H., 11, 95 n.14
Schwartz, J., 100 n.58, 102 n.107
Scyths, 58
Segal, C., 102 n.114
Selem, A., 105 n.160
Serpents, winged, 47
Serval, 37
Sex (sexual pleasure), 40, 41, 44, 45,
    46, 85, 106 n.168. See also Hunt
    (hunters, hunting)
Seyrig, H., 105 n.160
She-bear(s). See Bear(s)
Shmueli, E., 108 n.41
Sikyon, 81, 82
Simon, E., 35–37, 100 n.66, n.69, n.71,
    n.73, n.74, n.76
Smith, C., 111 n.24
Smith, P., 10, 95 n.11, 112 n.62
Smith, R., 69
Smyrna, 77, 86
Socrates, 39, 101 n.98
Sokolowski, F., 112 n.39, 113 n.49, 114
    n.91, n.94, 116 n.123
Solon, 43
Sophocles, 11
Sparta, 31
Sperber, D., 10, 95 n.11, n.13
Sphinx, 54
Spices, 8, 39, 50, 51, 56, 60. See also
    Perfumes
Spit, 9, 69, 74, 75, 79, 82, 89, 91, 93,
    111 n.29
splánkhna, 75–77, 85, 112 n.41, 113
    n.62
Stoic school, 56
Structuralist (structuralism). See
    Myths, structural analysis of
Styx, 75
Subversion, 26, 34, 52, 62, 72
Sun, 50, 81
sussplankhneúontes, 77

Table of the Sun, 59
Tannery, P., 109 n.53

Tantalos, 54
Tarentum, 66
Tarsus, 38
Taurus, 39, 63
tékhnē, 78
teleía, 31, 32, 91
teleîn, 33
Telemakhos, 76, 77
Teleté, 74, 84, 112 n.35
télos, 32, 33, 99 n.46, n.47, n.48
télos d'égnō, 32
Tenedos, 62
Tereus, 54, 55
Theban, 32
Theodota, 39
Thesmophoria, 14, 22, 103 n.127, 104
    n.155, 106 n.168
thesmophóroi, 14
Thrace, 92, 110 n.24
thrênos, 6
Thucydides, 16, 17
Thyestes, 53, 54, 112 n.38
Tierney, M., 115 n.107
títakos, 80
Titán, 80, 81
Titanē, 81
Titanìs gē, 80
títanos, 80, 81, 113 n.70
Titanos, 81
Titans, 8, 9, 54, 61, 62, 64, 69, 72–74,
    78–89, 91–94, 113 n.70
Titênios, 80
Todd, H. A. 101 n.91
Top, 69, 73, 111 n.27
Torre Nova, 92, 117 n.138
tò thēriôdes, 58
Toynbee, T.M.C., 102 n.101
Tragedy, 10, 12
Transgressions, 25, 52, 68. See also
    Subversion
tríbōn, 109 n.52
Trumpf, J., 102 n.115
Tsimshian, 19
Turan, 27
túrannos, 58
Turcan, R., 105 n.160, 115 n.101
Tylor, E., 53
Tyrant (tyranny), 58–59

Veii, 75
Vegetarians (vegetarianism), 56, 60, 61,
    66, 72
Vernant, J.-P., 12, 15, 95 n.15, 96 n.19,
    107 n.4, n.9, 108 n.16, n.25, n.27,
    114 n.83, 116 n.114, n.115, n.119
Veyne, P., 96 n.10
Vickers, B., 18, 19, 96 n.23

Vidal-Naquet, P., 95 n.15, 97 n.20,
    102 n.104, 107 n.9, 108 n.25, 116
    n.126
Vincigliata, 33
Virgins, 43
Viscera, 75–77, 85, 87. See also
    splánkhna
Voelke, A., 105 n.168, 107 n.10
von Fritz, K., 108 n.48
von Gaertringen, H., 107 n.6
Vulci, 37
Vultures, 50

Waltzing, J. P., 108 n.29
Welcker, 107 n.6
Wide, S., 99 n.52
Wilamowitz, 110 n.7, 114 n.84, 117
    n.139
Winckelmann, 18
Wine, 39, 76, 90

Wolves, 27, 44, 90
Woman, 92, 106 n.178, 117 n.138
Woodpecker, 55
Wool, 72
Wotke, F., 100 n.79

xenismós, 68
Xenophon, 24

Zancani-Montuoro, P., 103 n.123
Zeus, 32, 42, 44, 54, 69–71, 75, 80, 81,
    84, 85, 88
Zeus Conquerer, 45
Zeus Meilíkhios, 76
zōné, 99 n.51
zōsaménē, 32
zōstér, 99 n.53. See also Girdle
zōstēría, 32
Zumthor, P., 95 n.8
Zuntz, G., 116 n.131

Index Locorum

Aelianus, *De Natura Animalium* 5, 40: 38 (n.89); 5, 54: 38 (n.83); 6, 2: 37 (n.80)
———, *Varia Historia* 3, 42: 63 (n.37); 91 (n.128); 9, 17: 115 n.98; 13, 1: 31 (n.43)
Aeschylus, *Eumenides* 835: 99 n.47
———, *Prometheus* 172: 36 (n.70)
———, *Suppliants* 287: 33 (n.55)
———, fr. 84 Mette: 92 (n.133)
Aesop, *Fabulae* 42: 38 (n.82)
Alexis, *Pythagorizusa* frs. 2 and 3, *FCG* 3, 474–75 Meineke: 66 (n.50)
———, *Tarentini* frs. 1 and 2, *FCG* 3, 483 Meineke: 66 (n.56)
Ammianus Marcellinus, 22, 9, 15: 104 n.160; 19, 1, 11: 105 n.160
Anecdota Graeca I, 444, 30–445, 13 Bekker: 31 (n.43)
Anon., in *Comment. in Aristot. graeca* 20: 427, 38–40: 58 (n.24)
[Antigone], *Mirabilia* 31 in *Paradox. Graec.* 50–51 Giannini: 101 n.87
Antiphanes, *Mnemata*, *FCG* 3, 87 Meineke: 66 (nn.51 and 52)
Antoninus Liberalis, *Metamorphoses* 1: 44 (n.132); 10: 91 (n.128), 108 n.36
[Apollodorus], 1, 401, 2: 103 n.143; 2, 5, 9: 33 (n.53); 3, 9, 2: 31 (n.44), 33 (nn.57 and 58), 41 (n.103), 42 (n.110), 45 (n.137); 3, 14, 4: 35 (n.64)
Aratos, *Phainomena* 131: 111 n.29
Archilochus, frs. 168–74 Lasserre and Bonnard: 107 n.13.
Aristophanes, *Birds* 686: 81 (n.79)
———, *Frogs* 1032: 70 (n.11); 72 (n.18), 112 n.35
———, *Lysistrata* 785–96: 41 (n.106); 1014–15: 39 (n.96)
———, *Peace* 960: 73 (n.26); 1115: 113 n.52
———, fr. 478 Kock: 39 (n.96)
Aristophon, *Pythagoristes* fr. 1, *FCG* 3, 360–61 Meineke: 66 (n.50); fr. 3, *FCG* 3, 362 Meineke: 66 (n.50); frs. 4 and 5, *FCG* 3, 362–63 Meineke: 66 (n.52)
Aristotle, *De Generatione Animalium* 734d 16: 87 (n.114); 750a 30–35: 49 (n.152); 760b 23: 104 n.152; 774a 32–35: 104 n.148

———, *De partibus animalium* 666a 7–10: 87 (n.112), 115n. 111; 666a 20–22: 115 n.111; 666b 6–10: 87 (n.112); 667b 1ff.: 75 (n.41); 670a 23–26: 87 (n.110); 673b 1–3: 75 (n.43); 673b 15ff.: 75 (n.41); 674a 4–6: 75 (n.42)
———, *Historia Animalium* 6, 31, 579b 1–5: 48 (n.147); 9, 6, 612a 5–12: 101 n.86; 9, 6, 612a 12–15: 38 (n.87); 9, 614a 26–28: 40 (n.100)
———, *Meteorologica* 4, 3, 380a 36ff.: 78 (n.55); 381a 27ff.: 78 (n.55); 4, 11, 389a 28: 114 n.75
———, *Nichomachean Ethics* 6, 1148b 19–25: 58 (n.22)
———, *Politics* 1, 8, 1256b 7–26: 56 (n.10); 8, 1338b 19–22: 58 (n.22)
———, fr. 194 Rose: 85 (n.98)
[Aristotle], *Mirabilia* in *Paradox. Graeci* 226 Giannini: 101 n.86
———, *Problems* 3, 43 Bussemaker 4, 331, 15ff.: 69 (n.9), 74 (n.34); 5, 34, 884a 3ff.: 78 (n.55); 13, 4, 907b 35–37: 38 (n.85)
Athenaeus, 3, 84 C: 42 (n.118); 163 E-F: 66 (nn.54 and 55); 9, 410 A-B: 113 n.53; 14, 660 E: 75 (n.46)
Athenion *apud* Athenaeus 14, 660 E (=*CGF* 3, 369 Kock): 75 (n.46)
Aulus Gellius, 4, 11: 115 n.98

[Bion], *Lament for Adonis* 65–66: 50 (n.163)

Callimachus, *Hymn to Apollo* 85: 99 n.50
———, frs. 67–75 Preiffer: 44 (n.132)
Clement of Alexandria, 2, 19, 3: 103 n.127

Damascius, *Vita Isidori* 97: 39 (n.95)
Dio Chrysostom, 6, 21–22 Budé 1, 111, 4–11: 64 (n.43); 6, 25 Budé 1, 111, 23–28: 64 (nn.44 and 45); 6, 62 Budé 1, 120, 16–21: 64 (n.43); 10, 29–30 Budé 1, 145, 12–16: 65 (n.46); 30, 10: 83 (n.87); 30, 55: 80 (n.64)
Diodorus, 5, 73, 2–3: 44 (n.134)

Diogenes Laertius, 6, 56 and 105: 64 (n.43); 8, 28: 115 n.100
Dionysius Periegeta, 5, 935–47: 90 (n.125)
Dioscorides, 2, 173, 3: 104 n.160; 2, 176, 3: 50 (n.160)

Ephoros, *FGrH* 70, fr. 42 Jacoby: 59 (n.20); fr. 110: 20 (n.2)
Epicurus, *Rar. Sent.* 32 Usener: 56 (n.12)
Etymologicum Magnum s.v. gamēlía: 44 (n.134)
Euripides, *Bacchae* 702–68: 90 (n.126), 91; 1185–89: 62 (n.34); 1240–42: 62 (n.35); 1674: 63 (n.38)
————, *Cyclops* 243–47: 74 (n.38)
————, *Herakles* 416–18: 99 n.53
————, *Hippolytus* 17: 25 (n.23); 742–50: 103 n.121; 952–53: 70 (n.11); 954: 116 n.132
Eustathius, 332, 24ff.: 81 (n.77)

Firmicus Maternus, *De errore profanarum religionum* 6, p.15, 2 Ziegler: 72 (n.21)

Galen, 11, 831: 105 n.168

Herodotus, 2, 41: 111 n.29; 2, 64: 45 (n.139); 2, 81: 72 (n.17); 2, 97: 116 n.124; 2, 146: 116 n.124; 3, 25: 59 (n.27); 3, 108: 48 (n.146); 3, 108–9: 48 (n.145); 3, 110: 90 (n.124); 4, 106: 58 (n.21)
Hesiod, *Theogony* 459–60: 54 (n.3)
————, *Works and Days* 61: 81 (n.79); 72: 32 (n.49); 276–78: 57 (n.3)
————, fr. 76 Merkelbach-West: 32 (n.49), 41 (nn. 107, 108, and 109)
Hesychios *s.v. anemónē*: 104 n.158; *aphrodisía ágra*: 40 (n.99); *kardioústhai*: 114 n.92
Homer, *Iliad* 2, 189: 100 n.56; 2, 735: 81 (n.76); 6, 186: 100 n.56; 21, 573–80: 100 n.81
————, *Odyssey* 3, 5–66: 76, 77 (n.51); 11, 245: 32 (n.51); 11, 465: 104 n.158; 20, 74: 99 n.47
Homeric Hymn to Aphrodite 69–74: 44 (n.133)
Homeric Hymn to Demeter 372–74 and 411–13: 43 (n.128)

Hyginus, *Fabulae* 185: 33 (n.58), 46 (n.141); 185, 6: 45 (n.137)

Iamblichus, *De Vita Pythag.* 109 Deubner 63, 2–3: 115 n.98; 154 Deubner 87, 6–7: 79 (n.60)
————, *Protrepticus* 21, 108, 5–6 Pistelli: 85 (n.98)
[Iamblichus], *Theolog. arithm.* 22: 115 n.99
Ibykos, *P.M.G.* 286 Page: 43 (n.122)
Istros, *FGrH* 334 fr. 1 Jacoby: 81 (n.73)

Julian, *Disc.* 7, 214 C: 64 (n.43)

Lucian, *De Sacrificiis* 13: 114 n.92
————, *Lexiphanes* 23: 51 (n.170), 104 n.158

Menander, *Dúskolos* 456ff. and 519: 112 n.37
Mythographi Vaticani 1, 39: 46 (n.141), 103 n.136; 2, 47: 103 n.136

Nicander, frs. 65 and 120 Schneider: 51 (n.172)
Nonnos, *Dionysiaca* 6, 172–73: 73 (n.28); 12, 87–89: 103 n.137; 35, 82: 33 (n.56); 41, 155–57: 98 n.26; 41, 209–11: 98 n.26; 42, 209–11: 25 (n.65)

Oppian, *Cynegetica* 1, 24ff.: 98 n.26; 4, 320–53: 39 (n.94)
Orphicorum Fragmenta (Kern) 26: 87 (n.114); 31: 73 (n.27); 34: 112 n.35; 34–36: 73 (n.31), 110 n.8; 35: 69 (n.10), 74 (n.32), 74 (n.36), 114 n.88, 117 n.140, 117 n.141; 36: 114 n.88; 61: 88 (n.116); 85: 88 (n.116); 107: 88 (n.116), 114 n.82; 170: 88 (n.116); 207: 88 (n.117); 209: 73 (n.28), 73 (n.29), 80 (n.63); 209–14: 110 n.8; 210: 114 n.86, 114 n.89, 117 n.141; 215: 72 (n.21); 220: 114 n.87; 234: 117 n.138; 237: 88 (n.116); 291: 115 n.96; 301: 112 n.35
Orphicorum Fragmenta Testimonia (Kern): 194: 81 (n.81); 212: 70 (n.11); 213: 70 (n.11); 219: 115 n.96; 223: 112 n.35
Ovid, *Fasti* 3, 805ff.: 75 (n.44)

————, *Metamorphoses* 1, 228–29: 74 (n.38); 4, 190–255: 50 (n.155); 4, 250: 50 (n.156); 4, 256–70: 104 n.155; 4, 732: 104 n.156; 8, 317–430: 31 (n.37); 8, 322–23: 31 (n.38); 10, 520–739: 26 (n.28); 10, 537–41: 27 (n.29); 10, 547–52: 103 n.140; 10, 557–59: 34 (n.62); 10, 564–66: 30 (n.35); 10, 579: 106 n.178; 10, 609–80: 42 (n.111); 10, 637: 42 (n.112); 10, 644–48: 42 (n.116); 10, 676–78: 42 (n.113); 10, 681–82: 45 (n.135); 10, 694: 45 (n.138); 10, 698–704: 41 (n.102); 10, 698–707: 45 (nn.136 and 140); 10, 725–39: 50 (n.153); 10, 728–31: 50 (n.154); 10, 732: 50 (n.156); 10, 732–39: 50 (n.157)

Panyassis, fr. 25k (b) Matthews: 98 n.33
Parmenides, 8, 4 and 8, 32; 42: 33 (n.54)
Paroemiographi Graeci (Leutsch-Schneidewin) 2, 150, 5: 40 (n.99); 2, 513, 5–8: 32 (n.52); 2, 770: 103 n. 122; 3, 635, 1–2: 50 (n.164)
Pausanias, 1, 37, 4: 115 n.96; 2, 11, 5: 81 (n.74); 2, 17, 4: 42 (n.119); 2, 32, 2: 99 n.52; 2, 32, 7–9: 44 (n.134); 3, 9, 2: 31 (n.42); 3, 12, 4ff.: 33 (n.60); 3, 13, 6: 33 (n.60); 5, 16, 2: 31 (n.41); 7, 37, 5: 81 (n.81); 7, 37, 8: 73 (n.30); 9, 17, 3: 32 (n.50); 10, 4, 3: 81 (n.79)
Pherekydes *apud FGrH* 3 fr. 16 Jacoby: 43 (n.121)
Philetas, fr. 18 Powell: 102 n.117
Philo, *Legum allegoriae* 2, 6 Montdésert 107: 86, 87 (n.109)
Philokhoros, *FGrH* 328, fr. 74 Jacoby: 80 (n.72); fr. 173: 113 n.56
Philolaos, B 17 Diels-Kranz⁷: 87 (n.113)
Philostratus, *Imagines* 1, 6, 5: 48 (n.149)
————, *Life of Apollonius of Tyana* 2, 1–2: 39 (n.92); 4, 28: 42 (n.120)
Photius, *Bibliotheca* 324a 25–b 18: 39 (n.92)
Physiologus 1, 16 Sbordone p.16, 5ff.: 38 (n.90)
Pindar, *Pythian Odes* 9, 105–24: 34 (n.61)
Plato, *Euthydemus* 301a: 74 (n.33)
————, *Laws* 782b 6–c 2: 58 (n.23); 782c: 71 (n.15), 70 (n.11)

————, *Politicus* 271d: 57 (n.13)
————, *Protagoras* 320d: 81 (n.80); 321a: 57 (n.13)
————, *Republic* 364e: 91 (n.132); 372d–373a: 78 (n.57); 404a–d: 78 (n.58); 571c: 59 (n.26)
————, *Symposium* 196b: 51 (n.171)
————, *Timaeus* 73a: 112 n.41
Pliny, *Historia Naturalis* 8, 43: 47 (n.143); 8, 62: 38 (nn. 84 and 88); 13, 6: 38 (n.86); 21, 39: 38 (n.84); 21, 64: 50 (n.161); 21, 165: 104 n.156, 105 n.168
Plutarch, *Aqua an ignis util.* 2, 956 B: 64 (nn. 44 and 45)
————, *Camillus* 5, 5–6: 75 (n.45)
————, *Conjugal precepts* 138 D: 43 (n.126)
————, *De def. or.* 435 C and 437 A–B: 73 (n.26)
————, *De Esu Carnium* 995 C–D: 64 (n.42); 996 C: 83 (n.86); 998 A: 111 n.29
————, *Moralia* 500 C–D: 38 (n.82)
————, *Quaestiones Conviviales* 2, 3, 1, 635e–f: 86 (n.99); 729f: 73 (n.26)
————, *Quaestiones Graecae* 38, 299 E: 62 (n.36); 299 E–F: 91 (n.128); 299 E–300 A: 89 (n.121)
————, *Quaestiones Romanae* 264 B: 44 (n.134)
————, *Symposium* 3, 1, 646b: 50 (n.175); 648a: 50 (n.165)
Pollux, 3, 38: 44 (n.134)
Porphyry, *De Abstinentia* 1, 4: 56 (n.11); 1, 6: 56 (n.10); 1, 13: 57 (n.14); 2, 8: 63 (n.39), 89 (n.120), 92 (n.137); 2, 9: 73 (n.26); 2, 57: 59 (n.28)
————, *Perì agalmátōn* 7, p.10 Bidez: 105 n.160

Scholia AD in Il. 1, 449: 112 n.47
Scholia in Apollon. Rhod. 1, 425: 73 (n.26)
Scholia in Aristoph. Nubes 996–97: 44 (n.130)
Scholia in Aristoph. Peace 960: 73 (n.26)
Scholia in Euripid. Hippol. 1421: 35 (n.64)
Scholia in Euripid. Phoeniss. 150: 30 (n.36)
Scholia Lycophr. 208, p.98, 5 Scheer: 117 n.141; 831, p.266 Scheer: 35 (nn.63 and 65), 50 (n.162)

Scholia Theocr. (Wendel) 3, 38b: 42
  (n.117); 3, 40d: 30 (n.36); 5, 92e:
  50 (n.167)
Scholia T in Il. 24, 31: 26 (n.27)
Scholia vetera in Nic. Alex. (Keil) 375:
  104 n.154
Seneca, Thyestes 1060–65: 74 (n.38)
Servius, In Verg. Aen. 3, 13: 42
  (n.117); 3, 113: 46 (n.141)
———, In Ecl. 6, 61: 42 (n.117)
Solon, Nomoi frs. 127 a, b, c
  Ruschenbusch: 43 (n.125)
Stesichorus, P.M.G. 187 Page: 43
  (n.124)
Strabo, 8, 3, 14: 104 n.154; 10, 483 C:
  25 (n.22)

Theognis, 1275–77: 52 (n.177); 1289
  and 1294: 32 (n.48); 1289–94: 32
  (n.45)
Theophilus, Ad Autolycum 3, 5: 65
  (n.47)
Theophrastus, De Causis Plantarum
  6, 5, 1: 50 (n.166); 6, 5, 2: 101 n.87;
  6, 17, 9: 38 (n.84)
———, Historia Plantarum 6, 8, 1–2:
  50 (n.161) apud Scholia AD in Il.
  1, 449: 112 n.47
Thucydides, 3, 94: 58 (n.19)
Timotheus of Gaza, 11 Haupt: 39
  (n.94)

Vergil, Eclogue IV: 106 n.174

Theocritus, 3, 40–42: 42 (n.114); 5,
  88: 44 (n.131); 5, 92: 50 (n.164)
———, Epithalamion for Helen: 31
  (n.40)

Xenophanes B 33 Diels-Kranz[7]: 81
  (n.79)
Xenophon, Memorabilia 3, 1, 1, 5f.: 40
  (n.98)

Library of Congress Cataloging in Publication Data

Detienne, Marcel
  Dionysos slain.

  Translation of Dionysos mis à mort.
  Includes bibliographical references and indexes.
  1. Dionysus.   2. Myth.   I. Title.
BL820.B2D4613   292'.2'11   78-20518
ISBN 0-8018-2210-6